International Stratigraphic Guide

INTERNATIONAL STRATIGRAPHIC GUIDE

A GUIDE TO STRATIGRAPHIC CLASSIFICATION, TERMINOLOGY, AND PROCEDURE

by
International Subcommission on Stratigraphic
Classification of IUGS Commission on Stratigraphy

Hollis D. Hedberg, Editor

Editorial Committee: Neville George, Hollis Hedberg,
Charles Pomerol, Amos Salvador, and Jovan Stöcklin

A Wiley-Interscience Publication

JOHN WILEY AND SONS New York • London • Sydney • Toronto

Library of Congress Cataloging in Publications Data

International Union of Geological Sciences. Inter-
    national Subcommission on Stratigraphic Classifi-
    cation.
    International stratigraphic guide.

    "A Wiley-Interscience publication."
    Bibliography: p.
    Includes index.
    1. Geology, Stratigraphic. I. Hedberg, Hollis
Dow, 1903-   II. Title.

QE651.I57 1976   551.7   75-33086
ISBN 0-471-36743-5

Printed in the United States of America

10 9 8 7 6 5 4 3 2

Preface

Stratigraphy is a global subject, and international (global) communication and cooperation are necessary if we are to adequately comprehend the picture of the rock strata of the Earth as a whole, and to restore the history of *how, when,* and *why* these strata came to be *what* and *where* they are today.

Agreement on stratigraphic principles, terminology, and classificatory procedure is essential to attaining a common language of stratigraphy that will serve geologists worldwide. It will allow their efforts to be concentrated effectively on the many real scientific problems of stratigraphy, rather than being wastefully dissipated in futile argument and fruitless controversy arising because of discrepant basic principles, divergent usage of terms, and other unnecessary impediments to mutual understanding.

This *International Stratigraphic Guide* has been prepared by the International Subcommission on Stratigraphic Classification (ISSC) for those geologists anywhere (whether in academia, in research, or in industry) who wish to express their observations and thoughts on stratigraphy more clearly and, in turn, who wish to comprehend more clearly the stratigraphic information presented by others. It is particularly aimed at the needs of those whose work and interests are worldwide, or, at least, take them across national boundaries.

Diverse as are the strata of the Earth and their properties, they are certainly no more diverse than are the natures and characters of the persons who study them. All of our classifications and terminologies of natural bodies are no more than an attempted ordering contrived by human beings for the purpose of aiding our own imperfect conception and understanding of the infinite complexities of nature; and as such they have all the weaknesses of the human minds in which they have originated. Classification and terminology of rock strata are no exception.

The process of bringing together the diverse views on stratigraphic principles, classification, and terminology, developed from many dif-

v

ferent backgrounds, and of building these into a single set of practical terms and guidelines to which almost all can in large measure subscribe, has necessarily been a slow and patience-consuming task. Perhaps only those who have lived through the 20 years of work with the Subcommission can appreciate the thorny path along which it has had to make its way, contending first with apathy and then at times with fierce opposition, and continually being forced to pick its way through thickets of nationalism, regionalism, traditionalism, conservatism, and radicalism, in trying to arrive at the best majority consensus. It is of course a task that never will be finished. The publication of the present *Guide* should be looked on merely as a step in what should be a continuing course of progress to meet the continually growing needs of the science.

The present first complete edition of the *Guide* reorganizes and consolidates the thought of the many previously published circulars and preliminary reports of the Subcommission into what is hoped to be a more coherent and comprehensive whole. It attempts to eliminate unnecessary repetition, clarify conclusions, demonstrate certain points graphically by figures, and in general improve organization, wording, and manner of expression. An extensive bibliography of some 1500 entries accompanies the *Guide* and in itself provides an important part of the history of the growth of ideas incorporated in the *Guide*. It is a source of reference to many divergent views and philosophies worthy of consideration, which for want of space could not be discussed in the *Guide* itself.

The preliminary consolidation of the earlier reports into a draft of a single-volume *Guide* was carried out in 1974 by a small editorial committee consisting of Neville George, Charles Pomerol, Amos Salvador (vice-chairman of the Subcommission), Jovan Stöcklin, and the chairman, who together have carried the chief burden of this task. Their draft was then sent to the other members of what was originally an 11-man editorial board selected jointly by V. V. Menner and H. D. Hedberg in 1972 (Ivo Chlupáč, H. K. Erben*, Martin Glaessner, Ian Speden, Ryuzu Toriyama, and A. I. Zhamoida), who supplied additional valuable criticism. Another draft was then referred to the full membership of the Subcommission (125 members), who offered further useful suggestions and criticism, and by a vote of 85 to 3 approved publication (see Appendix D). (It is emphasized that the affirmative votes, as in the case of previous ISSC Reports, indicate

* In a letter of September 26, 1974, Professor Erben asked to be disassociated from the editorial board.

approval for publication, but not necessarily full agreement with all of the substance of the text.)

Continuing work by the Subcommission will now be devoted to studies and recommendations on the many special aspects of stratigraphy to which adequate attention could not be given in this first unified edition of the *Guide*. It will also be devoted to consideration of criticism and suggestions that have been invited from geologists in general as background for the eventual preparation of improved further editions.

In conclusion, the editor expresses his appreciation to all, members and many non-members, who have contributed to the preparation of this volume, and expresses his hope that it may be of service to stratigraphers all over the world and to the geological public in general. He also acknowledges his gratitude to the International Commission on Stratigraphy, the International Union of Geological Sciences, and the International Geological Congresses for their support and help with the publication of this and earlier Reports.

Finally, particularly grateful acknowledgment is due to Dr. Amos Salvador for constant help and advice in the preparation and assembly of text and illustrations and for unremitting efforts to improve the *Guide*; to Mrs. E. J. Spencer for her willing and efficient work that went far beyond the "call of duty" in producing the numerous drafts of both text and bibliography that have been necessary; and to Frances Hedberg for assistance in innumerable ways throughout the course of the work and for her help in the checking and preparation of the material for submission to the publisher.

HOLLIS D. HEDBERG, EDITOR
Chairman of International Subcommission
on Stratigraphic Classification (ISSC)
118 Library Place, Princeton, N.J.   08540, USA

October 1975

Contents

FIGURES

TABLES

International Stratigraphic Guide

Chapter One
Introduction

A. ORIGIN AND PURPOSES OF THE GUIDE

This *International Stratigraphic Guide* has been prepared by the International Subcommission on Stratigraphic Classification (ISSC)* of the International Commission on Stratigraphy. The Subcommission was created by the 19th International Geological Congress (Algiers), 1952; and has worked, first, under the aegis of the International Geological Congresses, and then, since 1965, under the International Union of Geological Sciences (IUGS). The eventual production of an international stratigraphic guide has been a principal objective of the Subcommission.

The purposes of the *Guide* are to promote international agreement on principles of stratigraphic classification and to develop a common internationally acceptable stratigraphic terminology and rules of stratigraphic procedure—all in the interests of improved international communication, coordination, and understanding, and thus of improved effectiveness in stratigraphic work throughout the world.

The recommendations of this first edition of the *International Stratigraphic Guide* are based on the current consensus of members of the Subcommission. Future editions of the *Guide* will undoubtedly introduce changes dictated by the tests of time and usage and will need to evolve with worthy new views and methods. Many special or newly developing fields of stratigraphy, such as those dealing with electrical and other kinds of well logs, seismic stratigraphy, magnetic reversals, geochemical zonation, soils, volcanogenic strata, igneous and metamorphic rocks, unconformity-bounded units, eustatic cycles, oceanic stratigraphy, ecostratigraphy, the Quaternary, the Precambrian, are barely touched in the present edition. These will be the subject of future studies by the Subcommission to be presented in supplements or revisions to be issued from time to time.

The Subcommission welcomes criticism and alternative proposals,

---

* Originally International Subcommission on Stratigraphic Terminology (ISST).

and it invites suggestions from all geologists for improving this first edition of the *Guide*.

B. COMPOSITION OF SUBCOMMISSION

The membership of the Subcommission has evolved over a 20-year period, beginning in 1954 with invitations to some 300 members of the Commission on Stratigraphy to join the Subcommission. Twenty-five accepted, and with the addition of a few other stratigraphers, the charter membership of the Subcommission amounted to 32 individuals (listed by name in Appendix A as recorded March 7, 1955). Membership has since gradually grown to 125 (listed by name in Appendix A as of December 31, 1974). The present composition represents a worldwide geographic spread of stratigraphers and stratigraphic organizations, and a wide spectrum of stratigraphic interests, traditions, and philosophies. Three classes of members are recognized as follows:

Individual Members: 51 members from 27 countries. This group includes volunteers from the Commission on Stratigraphy; other individuals added with the approval of the Commission on Stratigraphy to obtain viewpoints from nations, regions, or branches of the profession not otherwise well represented; and still other individuals added to represent points of view differing from those of the majority of members.

Individual Members ex officio: 35 members from 17 countries. This group includes the President of the IUGS Commission on Stratigraphy, and the presidents or secretaries of each of the Subcommissions, regional committees, and working groups of that Commission.

Organizational Members: 39 members consisting of stratigraphic committees and commissions, geological societies, and geological surveys from 33 countries and 4 multinational regions—and including almost every national or multinational stratigraphic organization in the world.

With these three kinds of membership, the Subcommission is considered reasonably representative, especially since it also has frequently gone to stratigraphers beyond its membership to solicit views on special problems.

C. PREPARATION OF THE GUIDE

Because of the size and geographic spread of its membership, coming from more than 40 countries, the Subcommission, in preparing the *International Stratigraphic Guide,* has had to operate largely by correspondence, supplemented by meetings at all International Geological Congresses since the Subcommission was created in 1952. It has particularly used the procedure of written questionnaires, responses, discussion, and conclusions.

The logical first step toward preparation of the *Guide* was to explore and evaluate existing stratigraphic principles, procedures, and terminology worldwide, and to determine to what extent there could be general agreement, or possibilities of obtaining general agreement, leading to an optimum and universally acceptable working basis in the field of stratigraphy. The results of these initial and time-consuming efforts are recorded in the many hundreds of pages of questionnaires and answers constituting the Circulars of the Subcommission during the first 10 years of its existence. Well-attended open discussion meetings were held at the International Geological Congresses in Mexico (1956) and Copenhagen (1960). This early exploratory work toward an international guide culminated in the publication by the 21st International Geological Congress (Norden) in 1961 of the Subcommission's first report, *Principles of Stratigraphic Classification and Terminology.*

During the following years, interest in stratigraphic principles and in standardization of stratigraphic terminology became increasingly widespread. Many new national and regional stratigraphic codes appeared, and new subcommissions of the Commission on Stratigraphy were created to deal with the stratigraphic problems of individual systems. Meanwhile, the Subcommission continued to try to ascertain through numerous circulars and questionnaires the views of its members and others on basic points of stratigraphic principle and procedure and to extend areas of agreement. As a further step towards the *International Stratigraphic Guide,* and in response to recognition by the Commission on Stratigraphy of the pressing need for more uniformity in the approach to defining individual systems and their boundaries, the Subcommission issued its second report, *Definition of Geologic Systems,* published in 1964 by the 22nd International Geological Congress (India).

The preparation of specific draft chapters of the *Guide* began in 1967 and preliminary versions of chapters on lithostratigraphic units, on stratotypes, on biostratigraphic units, and on chronostratigraphic units were published as ISSC Reports 3, 4, 5, and 6, respectively, by the 24th International Geological Congress (Canada) in 1970 and 1971. An *Introduction* and *Summary* to the proposed *Guide* (ISSC Report 7) was published in the periodicals *Lethaia* and *Boreas* in 1972. From these preliminary reports, plus the more than 2000 pages of questionnaires, responses, comments, and discussion contained in the 47 ISSC Circulars issued to date, the present single-volume *International Stratigraphic Guide* has been prepared.

The *Guide* is accompanied by a *Bibliography* of published literature on stratigraphic classification and terminology of some 1500 titles. A *Glossary* and a *Translation of Stratigraphic Terms from English into Other Languages* are in course of preparation.

Appendix B gives bibliographic references to the published Reports of the Subcommission and lists the libraries worldwide where all of its Reports and Circulars are on file and may be consulted and from which reproduction copies generally may be obtained.

D. SPIRIT OF THE GUIDE

The Subcommission offers its *International Stratigraphic Guide* as a recommended approach to stratigraphic classification, terminology, and procedure—not as a "code." There is no intention that any individual, organization, or nation should feel constrained to follow it, or any part of it, unless convinced of its logic and value. As reiterated in prefaces to all of its reports, the Subcommission believes that matters of stratigraphic classification, terminology, and procedure should not be legislated. Real and lasting progress will be achieved only as geologists in general agree voluntarily on the validity and desirability of certain principles, procedures, and terms. The purpose of the *Guide* is to inform, to suggest, and to recommend; and it must continually evolve in keeping with the growth of geologic knowledge.

Because of the great diversity of stratigraphic views, the *Guide* has favored a broad and relatively unrestrictive approach in defining principles, proposing rules, and recommending procedures. Where two important lines of stratigraphic thought conflict, the *Guide* generally favors the less prohibitive one—the one that allows the greater freedom for both points of view. Where a clear-cut selection between views cannot be made on this or other bases, it has seemed usually desirable to give both views, if possible indicating preference between them rather than attempting a compromise position. Ultimately, conflicting views will probably be settled by the test of usage. Terms that are not useful will die out; procedures that are not useful will be abandoned. But it should be realized that leaving matters to the test of usage has its disadvantages, because satisfactory solutions may be attained only after decades of confusion.

Furthermore, the *Guide* recognizes that there are stratigraphic situations where hard and fast rules cannot be applied, and that often one must simply use common sense in deciding what in the long run will most effectively promote clarity, understanding, and progress.

The *Guide* reflects a belief that the ideal concepts behind stratigraphic classification and terminology should be maintained, even if in practice the ideal can only be approached and not perfectly achieved. A relaxation of standards rarely leads to progress; and if compromises must be made in practice they should be recognized for what they are.

Clarity of definition is as essential to stratigraphy as to any other science. The use of unreal concepts, vague and cloudy terms, and imprecise definitions is often defended because such procedure seems easier, simpler, more attractive, or more traditional than a more rigorous approach. However, if a concept or a term cannot stand attempts at precise definition, it is usually of dubious value.

E. NATIONAL AND REGIONAL STRATIGRAPHIC CODES

Although the ultimate goal of the Subcommission has been a single set of common worldwide principles and rules of stratigraphic classification, terminology, and procedure, rather than many diverse national and regional schemes, the Subcommission has always encouraged the development of national and regional stratigraphic codes. Those that have come to the attention of the Subcommission are listed in order of their date of publication in Appendix C. These codes have helped in developing principles, in promoting awareness of the need for rules, and in providing a testing ground for observing the usefulness and applicability of certain proposals. However, the time is perhaps approaching when nations should prefer to concentrate on improving a single body of international rules rather than on developing numerous more or less conflicting local or national codes.

F. ALTERNATIVE OR DISSENTING VIEWS

The Subcommission recognizes that among geologists worldwide there are many different views on stratigraphic classification and stratigraphic terminology. Many of these views have been discussed thoroughly during the last 20 years in the Circulars of the Subcommission and in each of the Subcommission's published Reports space has been given freely to dissenting comments. Many of the principal dissenting opinions were further summarized in ISSC Report 7a under the heading of "Some Controversial Questions."

Initially consideration was given to attempting to publish dissenting comments and alternative views in the present *Guide*, but by decision of the majority of the Editorial Committee this plan was abandoned. It was pointed out that it would be manifestly impossible to devote adequate space to include all possible qualifying or conflicting ideas (and the rebuttals to these), that it would be unfair to select some and neglect others, that summaries were commonly unsatisfactory and that dissenting views were best kept in the author's own words, that most such views had already been published independently, and that references to those that had been published would be contained in the bibliography accompanying the *Guide* where they would be available to all who were interested.

Despite the mentioned diversity of views on stratigraphic principles in the world today, the approach of the *Guide* is believed to be sufficiently broad and tolerant so that it will not be found unduly restrictive to anyone in its application to practical problems of stratigraphy, regardless of divergent theories espoused.

G. VOTE ON PUBLICATION OF THE GUIDE

In keeping with its customary policy, the Subcommission has polled its membership as to whether the *Guide* should be published in its present form (emphasizing that a vote in favor of publication did not necessarily indicate full agreement with all parts of the text). The results of this poll are given in Appendix D. There were 85 votes in favor and 3 votes opposed.* In all three classes of membership, opinion overwhelmingly favored publication.

---

* Although no attempt has been made to publish dissenting comments, for reasons indicated in Section I.F, the editor believes it may be appropriate and of interest to summarize the reasons given for the three negative votes. One person objected that the *Guide* did not sufficiently take into account practice in Europe and conditioned his approval for publication on either (1) putting the chapters on lithostratigraphic and biostratigraphic units in an appendix and adding the paper of Zhamoida and Menner (1974) as a new chapter, or (2) republishing with the *Guide* the paper by Laffitte et al. (1972). See Bibliography for these references. Another person raised similar objections, disapproved of the treatment of zones, objected to the recognition of lithostratigraphic and biostratigraphic units on the grounds that "only chronostratigraphic categories are considered as stratigraphic," and wanted to have generally accepted material differentiated from material not accepted by everybody by using a different kind of type. The third person objected to the significance given to the term "zone" and stated that the *Guide* contained too many "foregone conclusions and sophistries." Although all three negative votes were from Europe, it should be noted that there were at the same time 40 affirmative votes from Europe.

Chapter Two
Principles of
Stratigraphic Classification

A. GENERAL

The whole Earth is stratified, in a broad sense, so that all rocks and all classes of rocks—sedimentary, igneous, and metamorphic—fall within the scope of stratigraphy and of stratigraphic classification.

Rock strata have many different properties. It is possible to classify stratified rocks according to any of their properties: lithology, fossil content, magnetic polarity, electrical properties, seismic response, chemical or mineralogical composition, and many others. Rock strata can also be classified according to such attributes as their time of origin or their environment of genesis.

The stratigraphic position of change for any one property or attribute does not necessarily coincide with that for any other. Consequently, units based on one property do not generally coincide with units based on another, and their boundaries not uncommonly cut across each other. Therefore, it is not possible to express with a single kind of stratigraphic unit the stratigraphic changes in all different properties or attributes; a different set of units is needed for each (see Figure 1).

At the same time, the general unity of stratigraphy should be emphasized. While many different kinds of units are needed to express intelligibly the variations in all of the many different properties and attributes of rock strata, still, these units are all closely related. They concern only different aspects of the same rocks, and they are all involved intricately with each other in achieving the same major goals of stratigraphy—to improve our knowledge and understanding of the Earth's strata and from this to outline the nature of past events, processes, and life on Earth.

7

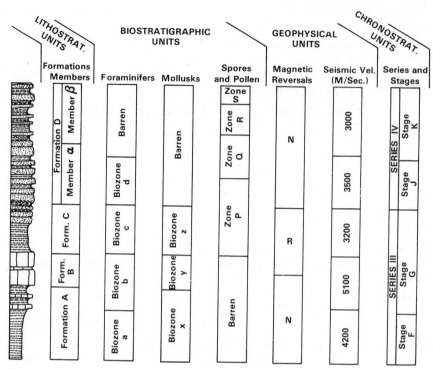

Figure 1. Illustration of differences in position in a stratigraphic section of division points for different properties or attributes of strata.

B.   CATEGORIES OF STRATIGRAPHIC CLASSIFICATION

Rock strata may be classified into many different categories, each of which needs its own distinctive units. The units of the following three categories are the best known and the most widely used:

1.   *Lithostratigraphy*—that element of stratigraphy which is concerned with the organization of strata into units based on their lithologic character.

2.   *Biostratigraphy*—that element of stratigraphy which is concerned with the organization of strata into units based on their fossil content.

3.   *Chronostratigraphy*—that element of stratigraphy which is concerned with the organization of strata into units based on their age relations.

Also widely used are units based on electrical properties, on seismic character, on heavy detrital minerals, and on magnetic polarity: and there are many others. No one can, nor need, use all possible kinds of stratigraphic units, but the way should be open to use any that promise to be useful; and it should be clear from the unit-terms to which category of classification any named unit belongs.

Though each kind of stratigraphic unit may be particularly useful in stratigraphic classification under certain conditions or in certain areas or for certain purposes, one kind—chronostratigraphic—offers the greatest promise for named units of worldwide application. Lithostratigraphic, biostratigraphic, and other similar kinds of stratigraphic units are restricted by the limited areal extent of the features chosen to characterize and distinguish them; and few, if any, of these features are both distinctive and present worldwide. Chronostratigraphic units, on the other hand, are based for definition on their time of deposition or formation—a universal property. In principle, they can be recognized the world over to the extent that the time-diagnostic features distinctive of the unit can be identified in the rocks.

Because chronostratigraphic units can often be recognized worldwide, they also offer the best means for international communication among stratigraphers with respect to position in the stratigraphic column. Any stratigrapher will readily understand if a colleague states that he has been studying the Jurassic, or the Miocene, or the Turonian of some area. However, if only the name of a formation, of a biostratigraphic zone, or of some other type of more local stratigraphic unit is mentioned, stratigraphers in other parts of the world may not be able to recognize even approximately the position of the unit within the stratigraphic column.

C. DISTINGUISHING TERMINOLOGIES FOR EACH CATEGORY

To realize the advantage of the stratigraphic tools provided by the various categories of stratigraphic units, appropriate distinguishing terminologies are needed for each. Elaborate terminologies have been developed through the years for the units of the most commonly used categories. For lithostratigraphy and chronostratigraphy the numerous terms represent different hierarchical ranks; for biostratigraphic units they result from the recognition of various kinds of biozones. For the units of newer or less used categories, only very simple terminologies (commonly zones of certain kinds) have been used thus far, but it may be expected that more elaborate schemes for some of these (e.g., magnetic polarity) may be developed before long. Table 1 gives terms here recommended for various stratigraphic categories.

Table 1. Summary of categories and unit-terms
in stratigraphic classification.[a]

Stratigraphic Categories	Principal Stratigraphic Unit-Terms	Equivalent Geochronologic Units
Lithostratigraphic	Group Formation Member Bed(s)	
Biostratigraphic	Biozones: Assemblage-zones Range-zones (various kinds) Acme-zones Interval-zones Other kinds of biozones	
Chronostratigraphic	Eonothem Erathem System Series Stage Chronozone	Eon Era Period Epoch Age Chron
Other stratigraphic categories (mineralogic, environmental, seismic, magnetic, etc.)	−Zone (with appropriate prefix)	

[a] If additional ranks are needed, prefixes <u>Sub</u> and <u>Super</u> may be used with
unit-terms when appropriate, although restraint is recommended to avoid com-
plicating the nomenclature unnecessarily.

D. CHRONOSTRATIGRAPHIC AND GEOCHRONOLOGIC UNITS

Each interval of stratified rocks represents a certain interval of geologic
time. Accordingly, each chronostratigraphic unit (interval of rock strata)
has a corresponding geochronologic unit (interval of geologic time).
Table 1 summarizes these units. Because geochronologic units are units
of geologic time—an intangible property—while stratigraphic units are

tangible material units composed of rock strata, geochronologic units are not in themselves stratigraphic units. To illustrate the difference, a chronostratigraphic unit can be likened to the sand that flows through an hourglass during a certain interval of time, while the corresponding geochronologic unit can be compared to the interval of time during which the sand flows. It may be said that the duration of the sand flow measures a certain interval of time—an hour, for instance—but the sand itself cannot be said to be an hour.

E. INCOMPLETENESS OF ROCK RECORD

Stratigraphic classification deals primarily with the Earth's *rock* sequence. However, it should be recognized that in any one area the rock record is far from continuous or complete. It is commonly broken by innumerable diastems, discontinuities, and erosional unconformities, and the evidence which it carries of these *missing intervals* is in itself a part of stratigraphy and a very important contribution to Earth history.

Chapter Three
Definitions
and Procedures

Certain definitions of general significance, and discussions of procedures pertinent to all kinds of stratigraphic units are assembled in this chapter to avoid unnecessary repetition in the chapters dealing with specific kinds of stratigraphic units.

A. DEFINITIONS

1. Stratigraphy, from the Latin *stratum* and the Greek *graphia,* is literally the descriptive science of strata and is used here simply as the *science of rock strata.* As such, stratigraphy is concerned not only with the original succession and age relations of rock strata, but also with their form, distribution, lithologic composition, fossil content, geophysical and geochemical properties—indeed, with all characters, properties, and attributes of rocks *as strata,* and their interpretation in terms of environment or mode of origin, and of geologic history. All classes of rocks—igneous and metamorphic as well as sedimentary, unconsolidated as well as consolidated—fall within the general scope of stratigraphy and stratigraphic classification. Some nonstratiform rock bodies are considered under stratigraphy because of their association with or close relation to rock strata.

2. Stratum. A geologic stratum is a layer (a generally tabular body) of rock characterized by certain unifying characters, properties, or attributes that distinguish it from adjacent layers. Adjacent strata may be separated by visible planes of bedding, or by less perceptible boundaries of change in lithology, mineralogy, fossil content, chemical constitution, physical properties, age, or any other property.

3. Stratigraphic Classification is the systematic organization of the Earth's rock strata, as they are found in their original sequence, into units with reference to any of the characters, properties, or attributes that rocks may possess. Many different properties and attributes of rock strata may serve usefully as the bases for stratigraphic classification, and so there are many different categories of stratigraphic classification.

4. Stratigraphic Unit. A stratigraphic unit is a stratum or assemblage of adjacent strata recognized as a unit (distinct entity) in the classification of the Earth's rock sequence, with respect to any of the many characters, properties, or attributes that rocks possess. Stratigraphic units based on one character will not necessarily coincide with those based on another; it is therefore essential that different terms be used for each so that their named units can be distinguished from each other. Clear definition of a stratigraphic unit is of paramount importance.

5. Stratigraphic Terminology deals with the unit-terms used in stratigraphic classification, such as formation, stage, biozone. It can be either formal or informal.

a. Formal stratigraphic terminology uses unit-terms that are defined and named according to an established or conventionally agreed scheme of classification, for example, the Chonta Formation, the Cretaceous System. The initial letter of a named formal unit term is capitalized.*

b. Informal stratigraphic terminology uses unit-terms only as ordinary nouns without the unit necessarily being named and without it being a part of a specific scheme of stratigraphic classification, for example, a chalky formation, an oyster zone. The initial letter of an informal unit-term is printed in lower case.

6. Stratigraphic Nomenclature deals with the proper names given to specific stratigraphic units, for example, Trenton Formation, Jurassic System, *Dibunophyllum* Range-zone.

---

* Recommendations in the *Guide* for the capitalization of terms are made with reference to the English language. It is recognized that these may not be applicable for use in certain languages with different rules of orthography.

7.   Stratotypes. Many kinds of stratigraphic units are best defined by reference to a designated type in a specific sequence of rock strata. A stratotype is the original, or subsequently designated, type representative of a named stratigraphic unit or of a stratigraphic boundary, identified as a specific interval or as a specific point in a specific sequence of rock strata, and constituting the standard for the definition and recognition of that stratigraphic unit or boundary (see Chapter 4 on Stratotypes).

8.   Zone. The term zone is commonly used for a minor stratigraphic interval in any category of stratigraphic classification. Thus there are many kinds of zones depending on the stratigraphic characters under consideration—lithozones, biozones, chronozones, mineral zones, metamorphic zones, zones of reversed magnetic polarity, and so on. When used formally, the term zone is given an initial capital letter (Zone) to distinguish it from its informal use.

9.   Interval. A stratigraphic interval is the body of strata between two stratigraphic markers. A geochronologic interval is the time span between two geologic events.

10.   Horizon. A stratigraphic horizon is an interface indicative of a particular position in a stratigraphic sequence. In practice it is commonly a distinctive, very thin bed. (The terms level, datum, marker, marker-bed, and key-bed frequently have been used in a similar sense.) There may be many kinds of stratigraphic horizons depending on the stratigraphic characters involved—lithohorizons, biohorizons, chronohorizons, seismic horizons, electrolog horizons, and so on. Among stratigraphic horizons may be included not only the boundaries of stratigraphic units, but also specific markers *within* these units that may be particularly useful for correlation purposes.

11.   Correlation. To correlate, in a stratigraphic sense, is to show correspondence in character and in stratigraphic position. There are different kinds of correlation depending on the feature to be emphasized. Lithologic correlation demonstrates correspondence in lithologic character and lithostratigraphic position; a correlation of two fossil-bearing beds demonstrates correspondence in their fossil content and in their biostratigraphic position; and chronocorrelation demonstrates correspondence in age and in chronostratigraphic position.

12.   Geochronology is the science of dating and determining the time sequence of events in the history of the Earth.

13.   A Geochronologic Unit is a subdivision of geologic time (time determined by geologic methods). It is therefore not a stratigraphic unit, although it may correspond to the time span of a stratigraphic unit (see Section 2D, p. 10-11).

14.   Geochronometry is that branch of geochronology that deals with the quantitative measurement of geologic time (usually in years).

15.   Facies, in stratigraphy, can mean *aspect, nature,* or *manifestation of character* (usually reflecting conditions of origin) of rock strata or specific constituents of rock strata. It is also used as a substantive for a body of rock strata distinctive in aspect, nature, or character. The general term "facies" has been greatly overworked. Rock strata may show differences in facies of various sorts so that one may speak of lithofacies, biofacies, mineralogic facies, marine facies, volcanic facies, boreal facies, and so on. If the term is used, it is desirable to make clear the specific kind of facies to which reference is made.

16.   Caution Against Preempting General Terms for Special Meanings. A source of much confusion and many of the controversies in stratigraphic terminology has been the preempting of general terms, with useful meanings in keeping with their etymology, for special restricted meanings. For example, "stratigraphy" should not be confined to *age* relations of strata; "correlation" is not necessarily *time*-correlation; "geochronology" should not refer exclusively to *isotopic* dating; "zone" can be applied to other than *fossil* zones; a "biozone" is not a specific kind of biostratigraphic zone; and "interval" may refer to either *time* or *space* intervals. A generally preferable procedure is to conserve the original general meaning of a term and to seek a more precise and less ambiguous word for the special meaning.

B.   PROCEDURES FOR ESTABLISHMENT AND DESCRIPTION OF STRATIGRAPHIC UNITS

A comprehensive definition of each named stratigraphic unit and a characterization of the type, of whatever kind, on which it is based are essential to its usefulness. Such definition should accompany the proposal for any new stratigraphic unit or the redefinition of any already existing unit. For lithostratigraphic and biostratigraphic units emphasis should be on lithologic and paleontologic characters respectively. For chronostratigraphic units, emphasis should be placed on features bearing on age and time-correlation. In general, a comprehensive definition and description of a stratigraphic unit should include the following topics as pertinent:

1. Name

Derivation of name; type locality.

2. Kind and Rank of Unit

Kind of stratigraphic unit; rank; general concept. New or revised definitions should include a statement of intent to introduce a new formal unit or to revise an already existing unit, and the reasons for doing so.

3. Historical Background

History of unit (author, original reference, previous treatment); synonymy; priorities; assurance against unnecessary duplication of already existing units.

4. Stratotypes and Other Standards of Reference

Geologic and geographic identification of stratotypes (verbal description, maps, structure sections, columnar sections, air photos, other photos, etc.). Provision for artificial markers. Relation of boundary-stratotypes to boundaries of other stratigraphic units and to other significant horizons in section (see Chapter 4 on Stratotypes).

For units of the type for which it is impractical to utilize stratotypes as standards, reliance will have to be placed entirely on the careful and accurate description and illustration of such features as do constitute appropriate standards of reference for the unit. Biostratigraphic units of this type particularly require clear description and figuring of diagnostic taxons, or literature references to such descriptions.

5. Description of Unit at Type Locality

Thickness, lithologic character, biostratigraphic character, structural attitude, geomorphic expression, unconformities or hiatuses, conditions of deposition, nature of boundaries of unit (sharp, transitional, un-conformable, etc.), and distinguishing and identifying features charac-terizing unit at type locality.

6. Regional Aspects

Geographic extent; regional variations in thickness, in lithostratigraph-ic, biostratigraphic, or other characters, or in geomorphic expression; regional stratigraphic relations; relations to other kinds of stratigraphic units, to marker horizons and so on; nature of boundaries away from type (sharp, transitional, unconformable, etc.); relations of boundaries of unit to boundaries of other stratigraphic units of same or different kind; criteria to be used in identifying and extending unit geo-graphically away from type locality.

7. Genesis (Where Appropriate)

Conditions of origin of rocks of unit; significance with respect to paleo-geography or geologic history.

8. Correlation with Other Units

9. Geologic Age

10. References to Literature

C. SPECIAL REQUIREMENTS FOR ESTABLISHMENT AND DESCRIPTION OF SUBSURFACE UNITS

Many useful local stratigraphic units are based on subsurface (well, mine, or tunnel) sections, and many more subsurface units will doubt-less be established as the sediments of oceanic areas become better explored. If adequate sample information is available, such subsurface sections can be used legitimately for establishing new stratigraphic units (Section 4.C.5, p. 29). The same general rules of procedure used for outcrop sections apply to subsurface units defined on the basis of exposures in mines or tunnels or from sections penetrated in wells. In proposing a name for a subsurface unit, the well or mine in which the type section is present becomes the type locality. In well sections, stratotypes need to be designated by well depths and on well logs rather than by markers at the surface, and geological information for these stratotypes will be based largely on well samples and well logs. Subsurface parastratotypes and hypostratotypes (see Section 4.A.6, p. 26) may be useful in supplementing poorly exposed surface stratotypes. The following data are desirable for the establishment and description of subsurface units:

1. Identification of Well or Mine

Name of type well or mine; location of well or mine by written description, map, exact geographic coordinates, farm or lease block, or any other geographic feature suitable for location identification; name of operating agency or individual. For mines: depth level where exposed. For wells: date of drilling, total depth, and surface elevation. If all the data needed to establish a type section cannot be furnished from one well, two or more wells should be used, with one being designated as the holostratotype and the others as parastratotypes or hypostratotypes.

2. Geologic Logs

Lithologic and paleontologic logs of the well or wells, and maps and cross sections of the mine, in written or graphic form, or both, are needed. The boundaries and subdivisions of a new unit should be indicated clearly on logs or charts.

3. Geophysical Logs and Profiles

Electrical or other mechanically recorded logs (preferably for several adjacent wells) and seismic profiles are very desirable. The boundaries and subdivisions of the unit should be marked clearly and should be shown at a scale large enough to permit full appreciation of detail.

4. Depositories

It is essential that sets of cuttings or other samples, fossil material, logs, and so on covering the type section of a stratigraphic unit be readily available for study. Such material should be placed at geological surveys, universities, museums, or other institutions with proper curatorial facilities. The location of the depository should be given.

If a named subsurface unit can be correlated with a named surface unit of the same kind, and if the characteristics of both are so similar that the two names are unnecessary, preference should generally be given to the surface unit as the type. But other factors should also be considered, such as priority of publication, usage, completeness of section, accessibility, nature of exposures in surface sections, and availability of type material from subsurface sections.

D. PUBLICATION

Establishment of a formal stratigraphic unit requires that a statement of intent and an adequate description of the unit be published in a recognized scientific medium. It is difficult to define strictly what constitutes "a recognized scientific medium" but the chief qualifications are scientific purpose and availability to the scientific public on request by purchase or through an accessible library. Regularly issued scientific journals meet this requirement. Many independent or irregularly issued publications also meet it, although in such cases some notice of the proposal should also appear in a widely circulated regularly issued scientific journal. Names proposed in informal or restricted media such as letters, company reports unavailable to the public, unpublished addresses, theses or dissertations, newspapers, and commercial or trade journals do not qualify. Publication of new stratigraphic names in abstracts issued in advance of complete reports usually does not establish these names, because the essential conciseness of abstracts does not permit adequate description. Casual mention or informal reference, such as "the formation at Jonesville schoolhouse" or "the limestones cropping out near San Francisco de Cara," does not establish a new formal unit, nor does mere use in a table or columnar section or on a map. To be valid, a new unit must be *duly proposed* and *duly described.*

E. PRIORITY

Priority in publication of a properly proposed, named, and defined unit should be respected. However, the critical factors should always be the usefulness of the unit, the adequacy of its description, freedom from ambiguity, and suitability for widespread application. Priority alone does not justify displacing a well-established name by one not well known or only occasionally used; nor should an inadequately established name be preserved merely on account of priority.

F. SYNONYMY

Before attempting to establish a new formal stratigraphic unit, authors should refer to national, state, or provincial records of stratigraphic names to determine whether a name has been previously used. The many volumes of the IUGS *Lexique Stratigraphique International* and other appropriate national or regional lexicons constitute valuable reference sources for most countries.

G. REVISION OR REDEFINITION OF PREVIOUSLY ESTABLISHED UNITS

Revision or redefinition of an adequately established unit without changing its name requires as much justification and the same kind of information as for establishing a new unit, and generally requires the same procedures. Redefinition may be justified to make a unit more useful or easier to recognize, map, and extend throughout the area of its development. Redefinition may also become desirable because of taxonomic changes of the fossil content of a biostratigraphic unit, or because of demonstrable and significant errors in an earlier definition. Names of long standing and common usage may be preserved legitimately if they are adequately defined even though their nomenclature may not conform to modern usage (see Section 5.F.3, p. 43).

H. SUBDIVISION

When a unit is divided into two or more units, the original name should not be employed for any of the subdivisions. The retention of the old name for one of the subdivisions would not only risk confusion but would also preclude use of the name in a term of higher rank.

J. CHANGE IN RANK

Change in rank of a stratigraphic unit does not require redefinition of the unit or its boundaries, or alteration of its proper name. Thus a stage may be raised to series rank or reduced to a substage, or a formation may be raised to a group or reduced to a member, without changing its name.

The rank of any stratigraphic unit should be changed only for substantial reasons and after careful consideration. Changes in major chronostratigraphic units of international scope should be made only after consultation with appropriate stratigraphic organizations.

K. REDUCTION IN NUMBER OF NAMES THROUGH CORRELATION

If correlation has established the identity of two named stratigraphic units, the later name should be replaced by the earlier, other considerations being equal, in the interest of simplicity in nomenclature.

L. UNCERTAINTY IN ASSIGNMENT

If there is uncertainty with respect to the assignment of strata to one or the other of two named units, it is always better to express this doubt rather than to make an arbitrary assignment. The following conventions may be used:

Devonian?	= doubtfully Devonian
Macoa? Formation	= doubtfully Macoa Formation
Peroc-Macoa formation	= strata intermediate in position (horizontally or vertically) between beds assuredly assigned to either of the two formations, which partake of the characters of both but which cannot be assigned decisively to either one and may eventually be made into a new formation
Silurian-Devonian	= one part Silurian and one part Devonian
Silurian or Devonian	= questionably either Silurian or Devonian
Silurian and Devonian (undifferentiated)	= both Silurian and Devonian but no distinction yet possible between the two

The name of the older or lower unit, if this distinction can be made, should always come first when two units are hyphenated.

M. ABANDONED NAMES

A name for a stratigraphic unit, once applied and then abandoned, preferably should not be revived except in its original sense. When it seems useful to refer to an obsolete or abandoned formal name, its status should be made clear by using such a phrase as "Mornas Sandstone of Hebert (1874)." To determine if a name has been abandoned or is obsolete, authors should refer to national, state, or provincial stratigraphic lexicons.

N. DUPLICATION OF NAMES

Duplication of names should be avoided. A name previously applied to any unit should not later be applied to another unless geographic separation precludes confusion.

P. RELATION OF NAMES TO POLITICAL BOUNDARIES

Stratigraphic units are not limited by international frontiers and effort should be made to use only a single name for each unit regardless of political boundaries. Spelling of a geographic name should generally conform to the usage of the country that contains the geographic locality from which the name has been taken.

Q. LINGUISTIC ROOTS OF UNIT-TERMS

A stratigraphic unit-term may be quite different in one language than in another (stage, étage, Stufe, piano, piso, etc.), or it may be very nearly the same in many languages (system, système, sistema, etc.). If a useful term is difficult to translate into a particular language, it may be desirable to "borrow" the term from the language of its origin, for example, range-zone from English. Stratigraphic terms with Greek or Latin roots are desirable because they are understood in a wide range of languages, for example, chronozone.

R. RECOMMENDED EDITORIAL PROCEDURES

(This edition of the *Guide* is written in the English language, and the editorial rules and procedures recommended here apply particularly to writing in that language. It is recognized that differing rules of orthography may make them inapplicable to writing in other languages.)

1. Capitalization

The unit-term of a formally named stratigraphic unit should always be capitalized, for example, *Bulimina-Bolivina* Assemblage-zone, Brunswick Formation, Upper Cretaceous Series, Devonian System. The use of capitals for formal unit-terms when these are not coupled with a proper name is discretionary, depending on needs for clarity or emphasis. Informal terms are not capitalized (except in those languages which require all nouns to be capitalized).

2. Hyphenation

Compound terms for most kinds of stratigraphic units, in which two common words are joined to give a special meaning, should be hyphenated, for example, range-zone, concurrent-range-zone. Exceptions are adjectival prefixes or combining forms which should generally be combined with the term-noun without a hyphen; for example, biozone, chronozone, subsystem, biohorizon, supergroup.

3. Clarity in Use of Zone Terms

The term "zone" is of value as a unit in many different categories of stratigraphic classification. However, the exact kind of zone should be clearly indicated, for example, biozone, chronozone, lithozone, range-zone, assemblage-zone, mineral-zone, zone of reversed magnetic polarity.

4. Names of Fossils

The printing of fossil names for stratigraphic units should be guided by the rules laid down in the *International Code of Zoological Nomenclature* and in the *International Code of Botanical Nomenclature.* The initial letter of generic names should be capitalized; the initial letter of species names should be in lower case; taxonomic names of genera and species should be in italics. Units named from a species should also give the genus name. After the first mention of the genus name it may be abbreviated to its initial letter if there is no danger of confusion with some other genus beginning with the same letter; for example, *Exus albus* may be shortened to *E. albus.*

Chapter Four
Stratotypes

A. DEFINITIONS

1. Stratotype (Type Section)—the original or subsequently designated type of a named stratigraphic unit or of a stratigraphic boundary, identified as a specific interval or a specific point in a specific sequence of rock strata, and constituting the standard for the definition and recognition of the stratigraphic unit or boundary.

2. Unit-Stratotype—the type section of strata serving as the standard for the definition and recognition of a stratigraphic unit. The upper and lower limits of a unit-stratotype are its boundary-stratotypes (see Figure 2*a*).

3. Boundary-Stratotype—a specific point in a specific sequence of rock strata that serves as the standard for definition and recognition of a stratigraphic boundary (see Figure 2*b*).

4. Composite-Stratotype—a unit-stratotype formed by the combination of several specified type intervals of strata known as *component-stratotypes*. Thus a certain lithostratigraphic unit may not be entirely exposed in any single section, and it may be necessary to designate one section as the type for the lower part of the unit and another section as the type for the upper part of the unit. In this case, one of the two component sections should be considered the holostratotype and the other a parastratotype.

A stratotype of a unit of higher rank formed by the combination of the stratotypes of its component units of lower rank is also a composite-stratotype. Thus the stratotype of a series may be a composite of the stratotypes of its constituent stages. In such a case, the lower boundary-stratotype of the lowermost constituent stage is also the

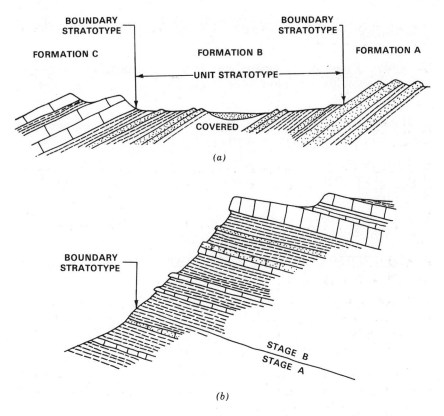

Figure 2. (a) Unit-stratotype and boundary-stratotypes for a lithostratigraphic unit (Formation B). (b) Boundary-stratotypes for chronostratigraphic units. (Upper boundary of Stage A is lower boundary of Stage B.)

boundary-stratotype of the base of the series. If the components of a composite-stratotype are already established formal stratigraphic units, it is unnecessary to distinguish one as a holostratotype and others as parastratotypes.

5. Type Locality and Type Area. The *type locality* of a stratigraphic unit, boundary, or other feature is the specific geographic locality in which its stratotype is situated, or lacking a designated stratotype, the locality where it was originally defined or named. The *type area* (or *type region*) is the geographic territory surrounding the type locality. A type locality or type area differs from a stratotype (type section) in that it refers to a geographic locality or area rather than to a specific profile or stratigraphic section.

6. Holo-, Para-, Neo-, Lecto- and Hypostratotypes. The following terminology adds precision to the designation and description of stratotypes:

a. Holostratotype—the original stratotype designated by the author at the time of establishment of a stratigraphic unit or boundary.

b. Parastratotype—a supplementary stratotype used in the original definition by the original author to aid in elucidating the holostratotype.

c. Lectostratotype—a stratotype selected later in the absence of an adequately designated original stratotype.

d. Neostratotype—a new stratotype selected to replace an older one which has been destroyed or nullified.

e. Hypostratotype (also called *reference section, auxiliary reference section*)—a stratotype designated to extend knowledge of the unit or boundary established by a stratotype to other geographical areas or to other facies. It is always subordinate to the holostratotype.

Thus a holostratotype and a parastratotype are originally designated primary types; a lectostratotype and a neostratotype are subsequently designated primary types; and a hypostratotype is a subsequently designated secondary (reference or auxiliary) type always subsidiary to a primary type.

Holostratotypes, parastratotypes, and lectostratotypes are generally situated within the type area. Neostratotypes, lectostratotypes, and hypostratotypes may be chosen beyond the limits of the original type area.

In principle, stratotypes should not be altered or amended. However, if an established stratotype is permanently destroyed, or if it has been found to have been established in violation of accepted stratigraphic principles, a new stratotype (neostratotype or lectostratotype) may be established, preferably in the type area. There may be more than one typical section but only one type section or stratotype.

B. STRATOTYPES IN DEFINITION OF STRATIGRAPHIC UNITS

1. Standard Definitions

Stratigraphy makes use of numerous named divisions of the stratigraphic column belonging to various stratigraphic categories—

lithostratigraphic units, biostratigraphic units, chronostratigraphic units, and so on. It is essential that these named units, and their boundaries, be clearly defined so that all who use them will start with the same basic understanding of their meaning and so that there will be a common standard for their recognition away from their places of origin. For many such units a designated stratotype provides an essential or at least useful aid to definition.

2. Reference to a Specific Rock Section

The concept of a stratigraphic unit is usually based on features or attributes of the rock strata—lithology, fossil content, age or time span, and so on—which are observable or verifiable in the rock strata and may be related in advance of naming to a specific interval of rock strata. The stratotype of such a unit, therefore, constitutes the standard of reference on which the concept of the unit is uniquely based. Stratigraphic units may be defined also by means of written descriptions; but, valuable as such descriptions are, they are always subject to misunderstanding due to differences in language, differences in interpretation of words, inadequacy or error in description, or other imperfections in our ability to convey concepts verbally. On the other hand, for many kinds of stratigraphic units and their boundaries, a designated and identified interval or point in a specified sequence of rock strata—a unit-stratotype or a boundary-stratotype—provides by far the most stable and unequivocal standard of definition.

(For some few kinds of stratigraphic units, such as biostratigraphic range-zones, the standard of the unit is a concept which cannot be tied in advance to a specified interval of strata since the stratigraphic scope of the unit may vary widely in the future with increasing information. Units of this sort cannot be defined satisfactorily by stratotypes.)

C. REQUIREMENTS FOR STRATOTYPES

Stratotypes for each different category of stratigraphic units—lithostratigraphic, biostratigraphic, chronostratigraphic—require individual consideration and are discussed in the chapters covering each type of unit. The following requirements apply to stratotypes in general.

1. Expression of Concept

The most important requisite of a stratotype is that it adequately represent the essentials of the concept for which it is the material type. A complete exposure of all strata in the unit from bottom to top and throughout its entire lateral extent would be the ideal stratotype. However, because it is impossible to find or establish such a comprehensive stratotype, reliance is usually placed on a single section, as complete and well exposed as possible. Lack of continuous exposures or the presence of structural complications may make it impossible to find even such a single continuous section through an entire stratigraphic unit. It may then be necessary to resort to a composite-stratotype or to supplementary and reference sections (parastratotypes and hypostratotypes), or to express the unit-stratotype simply as the stratigraphic interval between a designated boundary-stratotype marking the base of the unit and another designated boundary-stratotype marking the top of the unit.

In the case of chronostratigraphic units (e.g., stages), it is desirable that the lower (older) boundary-stratotype of one unit be the upper (younger) boundary-stratotype of the immediately underlying unit (Figure 2b, p. 25), thus avoiding difficulties in time-correlation which might leave gaps or overlaps between types (see also Figure 13, p. 85).

2. Description

The description of a stratotype should be both geographic and geologic. The geographic description should enable anyone to find the stratotype readily in the field. It should include a detailed map showing location and means of access to the type locality. It would also be desirable to include air photographs and other photographs at an appropriate scale to show the geographic extent of the unit in the type area and the geographic position of its boundaries.

The geologic description should cover thickness, lithology, paleontology, mineralogy, structure, geomorphic expression, and other geologic features of the type section. The boundaries and relations with adjacent units particularly should be described in detail, and reasons for choice of boundaries should be given. The description should be accompanied by graphic profiles, columnar sections, structure sections, and photographs.

3. Identification and Marking

An essential requirement of a stratotype is that it be clearly marked. A boundary-stratotype should be based on a single point in a designated sequence of rock strata, serving to indicate the position of the boundary horizon at one place. (Lateral extension of the boundary horizon in any direction from this point is accomplished by stratigraphic correlation.) A unit-stratotype desirably should be clearly delimited by boundary-stratotypes for its base and for its top. Preferably, a boundary-stratotype or the limits of a unit-stratotype should be indicated by a permanent artificial marker, but in any case boundary points should be described geographically and geologically in such detail that there can be no doubt as to their exact location.

4. Accessibility

If the stratotype is to fulfill the role of a standard, it should be situated in an area geographically accessible to all who are interested, regardless of political or other circumstances.

5. Subsurface Stratotypes

There is no objection to establishing subsurface stratotypes if adequate surface sections are lacking and if adequate subsurface samples and logs are available (see Section 3.C, p. 17).

6. Acceptability

Perhaps in no phase of stratigraphic classification is there greater need for worldwide collaboration than in setting up standards of definition for stratigraphic units of international extent so that they will be generally acceptable and so that geologists of all countries will use these units in the same sense. The usefulness of a stratotype is directly related to the extent to which it is generally accepted or acknowledged as *the type*. It is, therefore, always desirable, and in due course to be expected, that the designation of a stratotype be submitted for approval to the geological body having the highest standing in any particular case.

Stratotypes for chronostratigraphic units or boundaries of international or worldwide application should be approved by appropriate bodies of the highest international or worldwide geological standing. On the other hand, stratotypes of units of only local extent and interest may require approval from no more than local or national surveys or stratigraphic commissions.

Chapter Five
Lithostratigraphic Units

A. PURPOSE OF LITHOSTRATIGRAPHIC CLASSIFICATION

The purpose of lithostratigraphic classification is to organize systematically rock strata of the Earth into named units that will represent the principal variations of these rocks in lithologic character. All stratigraphic units are composed of *rock* and thus have "rock character," but only lithostratigraphic units are differentiated on the basis of *kind* of rock (lithologic character); limestone, sandstone, sand, tuff, claystone, basalt, marble, and so on. The recognition of such units is useful in visualizing the physical picture of the Earth's strata, in determining local and regional structure, in investigating and developing mineral resources, in determining the origin of rock strata, and in working out rock sequence.

Lithostratigraphic classification is usually a first approach in stratigraphic work in any new area. However, it also continues always to be an essential element of the stratigraphy of the area. Likewise, it is always an important key to geologic history. For instance, much can be deduced with certainty about the geologic events that have taken place in the area of Figure 3, even though no ages are available from either fossils or isotopic age determinations.

The relation of lithostratigraphic units to other kinds of stratigraphic units is discussed in Chapter 8.

B. DEFINITIONS

1. Lithostratigraphy—the element of stratigraphy that deals with the lithology of strata and with their organization into units based on lithologic character.

2. Lithostratigraphic classification—the organization of rock strata into units on the basis of their lithologic character.

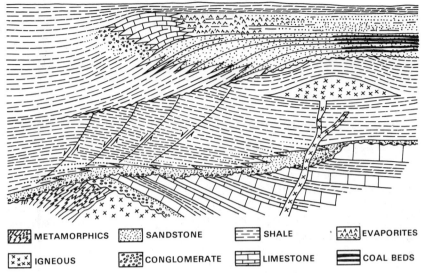

METAMORPHICS	SANDSTONE	SHALE	EVAPORITES
IGNEOUS	CONGLOMERATE	LIMESTONE	COAL BEDS

Figure 3. Significance of lithostratigraphic units in deciphering geologic history.

3. Lithostratigraphic unit—a body of rock strata that is unified by consisting dominantly of a certain lithologic type or combination of lithologic types, or by possessing other impressive and unifying lithologic features. A lithostratigraphic unit may consist of sedimentary, or igneous, or metamorphic rocks, or of an association of two or more of these. The rock may be consolidated or unconsolidated. The critical requirement of the unit is a substantial degree of overall lithologic homogeneity. (Diversity in detail may in itself constitute a form of overall lithologic unity.) Lithostratigraphic units are recognized and defined by observable physical features and not by inferred geologic history or mode of genesis. Fossils may be important in the recognition of a lithostratigraphic unit either as minor but distinctive physical constituents or because of their rock-forming character as in coquinas, diatomites, coal beds, and so on. The geographic extent of lithostratigraphic units is controlled entirely by the continuity and extent of their diagnostic lithologic features. Only major lithologic features readily recognizable in the field should serve as the basis for lithologic units.

4. Lithostratigraphic zone (lithozone)—an informal lithostratigraphic unit used to indicate a body of strata that is unified in a general way by lithologic features but for which there is insufficient need or insufficient information to justify its designation as a formal unit; for example, the shaly zone in the lower part of the Parker Formation, the coal-bearing zone exposed south of Ravar, the Burgan oil-producing zone.

5. Lithostratigraphic horizon (lithohorizon)—a surface of lithostratigraphic change or of distinctive lithostratigraphic character; preeminently valuable for correlation (not necessarily time-correlation): commonly the boundary of a lithostratigraphic unit, though also often a lithologically distinctive horizon or very thin marker bed within a lithostratigraphic unit.

C. KINDS OF LITHOSTRATIGRAPHIC UNITS

1. Hierarchy of Formal Lithostratigraphic Units

Formal lithostratigraphic units are those that are defined and named in accordance with an explicitly established or conventionally agreed scheme of classification and nomenclature (see Table 1, p. 10, and Section 3.A.5, p. 13). The conventional hierarchy of formal lithostratigraphic terms is as follows:

> *Group*—two or more formations
>> *Formation*—primary unit of lithostratigraphy
>>> *Member*—named lithologic entity within a formation
>>>> *Bed*—named distinctive layer in a member or formation

2. Formation

The *formation* is the primary formal unit of lithostratigraphic classification; it is a body of rock strata of intermediate rank in the hierarchy of lithostratigraphic units. Formations are the only formal lithostratigraphic units into which the stratigraphic column everywhere should be divided completely on the basis of lithology.

The degree of change in lithology required to justify the establishment of distinct formations (or other lithostratigraphic units) is not amenable to strict and uniform rules. It may vary with the complexity of the geology of a region and the detail needed to portray satisfactorily its rock framework and to work out its geologic history.

The thickness of units of formation rank follows no standard and may range from less than a meter to several thousand meters, depending on the size of units locally required to express best the lithologic development of a region.

Practicability of mapping and of delineation on cross sections is an important consideration in the establishment of formations.

Formations may be composed of sedimentary rocks, or extrusive or intrusive igneous rocks, or metamorphic rocks, or under some circumstances associations of two or more of these types of rock.

3. Member

A *member* is the formal lithostratigraphic unit next in rank below a formation and is always a part of a formation. It is recognized as a named entity within the formation because it possesses lithologic characters distinguishing it from adjacent parts of the formation. No fixed standard is required for the extent or thickness of a member.

A formation need not be divided into members unless a useful purpose is thus served. Some formations may be completely divided into members; others may have only certain parts designated as members; still others may have no members. A member may extend from one formation to another.

4. Bed

A *bed* is the smallest formal unit in the hierarchy of lithostratigraphic units. It is a unit layer in a stratified sequence of rocks which is lithologically distinguishable from other layers above and below, for example, the Baker Coal Bed. Several contiguous beds of similar lithology may also together constitute a formal unit, for example, the Marcus Limestone Beds.

The term "bed" is applied customarily to layers a centimeter to a few meters in thickness. Those of less thickness are called *laminae*. Stratification is not necessarily identical with splitting properties and the two should not be confused.

A single named bed or named sequence of beds may constitute a member or a formation, and in such a case, the unit term bed (or beds) is replaced by member or formation, for example, "Bracklesham Beds" = Bracklesham Formation, "Drusberg-Schichten" = Drusberg Formation. A specific bed conceivably may pass from one member or formation to another and still retain the same name.

Customarily only distinctive beds (commonly known as *marker beds*) particularly useful for stratigraphic purposes, like correlation or reference, are given proper names and considered formal lithostratigraphic units. Intervening beds are left unnamed.

In relatively unexplored country where only rapid reconnaissance can be carried out, a formal and properly surveyed hierarchy of formations and groups cannot always be established initially. In such cases, the use of a geographic name coupled with "beds" (uncapitalized) will be accepted as a temporary informal designation.

5. Group

A *group* is the formal lithostratigraphic unit next in rank above a formation. The term is applied most commonly to a sequence of two or more contiguous associated formations with significant unifying lithologic features in common. The type or reference sections of a group are the type or reference sections of its component formations. The proposal for recognition of a group should outline clearly the unifying characteristics on which it is based and the formations of which it is composed. Formations need not be aggregated into groups, but the term group is generally used for an assemblage of formations, except that it may also be used for a stratigraphic sequence of any kind of rocks that will probably be divided in whole or in part into formations in the future.

The aggregation of formations into groups provides a useful means of simplifying and generalizing stratigraphic classification for those who may not need, or may wish to avoid, the often complex detail of formational subdivision in certain regions or certain intervals.

The name of a group should preferably be derived from an appropriate geographic feature near the type areas of its component formations, for example, Sucre Group, Newark Group, Garhwal Group. Generally, it is not possible to characterize the lithology of a group in its name.

The component formations of a group are not necessarily everywhere the same. For example, in parts of western Iran the Fars Group comprises simply the Gach Saran and Agha Jari Formations; in parts of the province of Fars in southern Iran the group consists of the Razak, Mishan, and Agha Jari Formations. A formation may extend laterally from one group to another.

The wedging-out of a component formation or formations may justify the reduction of a group to formation rank, retaining the same name. A group may be extended laterally beyond the area where it is divided into formations, if it seems probable that the equivalent interval will eventually be divided into formations. Otherwise, the group becomes, in effect, a single formation and should be known as such.

6. Subgroup and Supergroup

A group may on occasion be divided into *subgroups*. The term *supergroup* may be used for several associated groups, or for associated formations and groups with significant features in common.

7. Complex

A *complex* is a lithostratigraphic unit composed of diverse types of any class or classes of rock (sedimentary, igneous, metamorphic) and characterized by highly complicated structure to the extent that the original sequence of the component rocks may be obscured. The term may be used as a part of a formal name in the place of a lithologic or rank term, for example, Akkajaure Complex, Lewisian Complex. The rank of a complex may be equivalent to a group, a formation, or a member. The term "basement complex" is applied informally to rocks of complicated structure (commonly igneous or metamorphic) underlying a less deformed, dominantly sedimentary sequence in any region.

8. Informal Lithostratigraphic Units

Informal lithostratigraphic units include lithologic bodies to which casual reference is made but for which there is insufficient need, insufficient information, or inappropriate basis to justify designation as a formal unit. These may be referred to informally as lithozones (e.g., shaly zones, coal-bearing zones) or as beds or members (lower case) of one sort or another (e.g., sandy beds, the pebbly bed, the shaly member). The term "measures" has been used informally for a sequence of coal beds (e.g., the Mara coal measures), but "The Coal Measures" has also been used historically for a series of the Carboniferous System.

When informal units such as those mentioned above are given geographic names, the unit-term should not be capitalized (unless some problem arises because of the languages involved). Geographic names should be combined with the terms "Formation" and "Group" only in formal nomenclature.

a. Industrial Units.   Lithostratigraphic bodies that are recognized more for ultilitarian purposes than for their lithologic unity, such as aquifers, oil sands, quarry layers, and orebearing "reefs," are considered to be informal units even if named.

b. Tongue and Lentil.   *Tongue* and *lentil* (lens, lenticle) have been used frequently as independent lithostratigraphic rank terms, but more correctly they are only specially shaped forms of members (or of formations). A lentil is a lens-shaped body of rock of different lithology than the unit that encloses it. A projecting part of a unit extending out beyond its main body may be called a tongue. The proper designation of these bodies rests in part on circumstances of exposure; a body seen as a lentil might be part of a narrow tongue. Tongues and lentils may be given names.

c. Reef.   *Reef* is a term commonly applied to a mass of organic carbonate built in place by corals, algae, or other sedentary organisms. Reefs are of variable size and shape and may occur as limited isolated masses or extensive continuous masses. A reef may constitute a member within a single formation, or it may extend upward or laterally through two or more formations, or it may constitute a formation in itself. Reefs may be given geographic names if they are of sufficient size or importance, for example, Thornton reef near Chicago, Illinois; Tratau reef near Sterlitamak, Bashkiria; and the subsurface Peace River reef and Grossmont reef complex in Alberta; but unless followed by the terms "Member" or "Formation," they are considered to be informal lithostratigraphic units.

d. Other.   Certain other rock bodies related to or closely associated with lithostratigraphic units are not true *litho*stratigraphic units because they emphasize mode of origin, or shape, or some character other than lithology for identification; others are not stratigraphic units at all. Among these two types are slides and slumps, mud flows, olistostromes, olistoliths, diapirs, salt plugs and walls, veins, batholiths, cyclothems, and such. These may be given informal names.

9.   Some Special Aspects of Igneous Rocks

Igneous rock bodies of more or less tabular form, conforming to the general stratification of the rock section, may constitute the basis for lithostratigraphic units, either in themselves or in combination with adjacent interbedded sedimentary strata of essentially the same age or much older.

Igneous rocks, however, present some special problems in lithostratigraphy. For example, many igneous rocks occur as dikes or other bodies injected *across* the dominant stratification of the section. Moreover, sills, flows, and other more or less accordantly interstratified intrusive or extrusive igneous rock bodies may have been emplaced in the stratigraphic section at a level far above that at which their parent magma originated, yet they may still be connected with the source body across intervening stratigraphic units by necks, pipes, or dikes of the same rock material. These cross-cutting igneous bodies do not in themselves constitute stratigraphic units but they are, of course, an important part of the lithologic picture. They may conveniently be referred to as "associated" with the strata of the lithostratigraphic units that they transect.

(A somewhat similar situation to that for igneous rocks obtains for intrusions or extrusions of sedimentary rock. However, there is an important difference. The intruding or extruding sediment, whether consolidated or unconsolidated, belonged to some already formed stratum previous to the time of its intrusion or extrusion, and therefore, where it did not form a new stratigraphic unit in itself, it may simply be referred to as a displaced part of the original source stratigraphic unit.)

D. PROCEDURES FOR ESTABLISHING LITHOSTRATIGRAPHIC UNITS

1. Stratotypes as Standards of Definition

Each named lithostratigraphic unit of whatever rank should have a clear and precise standard definition based on the fullest possible knowledge of its lateral and vertical variations. The designation of a type section (unit-stratotype) is essential in the definition of a lithostratigraphic unit.

The stratotype of a lithostratigraphic unit is the specifically designated sequence of rock strata on which the standard definition of the lithologic character of the unit is based. It is designated at a specific geographic locality, preferably that from which the name of the unit is derived.

The stratotypes of lithostratigraphic units of formation or lesser rank are commonly simple unit-stratotypes. The stratotypes of lithostratigraphic units larger than a formation (e.g., groups) are commonly composite-stratotypes, that is, a composite of the stratotypes of the component formations of the group.

Where strata are nearly horizontal or poorly exposed and a complete section of a unit does not crop out in a reasonably limited area, it may not always be practicable to designate any one specific complete and continuous section as the unit-stratotype. Only a type area or a type locality, rather than a type section, can be indicated. In such cases it is essential to identify explicitly the lower and upper boundary-stratotypes at specified sections where the change to underlying and overlying units can be seen. The equivalent of the unit-stratotype is therefore constituted by the ensemble of exposures in the type locality or type area lying stratigraphically between the lower and upper boundary-stratotypes.

Although the stratotype of a lithostratigraphic unit is a specific interval of strata in a specific section or area, the unit, as recognized elsewhere, may contain a greater or lesser thickness of strata than in the stratotype, and may span a greater or lesser time interval than the stratotype. The only critical requirement of the unit as identified elsewhere is that it have essentially the same lithology and relative stratigraphic position as in the stratotype.

2. Reference Sections (Hypostratotypes)

The definition of a lithostratigraphic unit based on the designation of a unit-stratotype section (holostratotype) is often usefully supplemented by designating in other geographic areas one or more auxiliary reference sections (hypostratotypes), sometimes better exposed or more accessible than the original unit-stratotype. These reference sections, however, must always be considered as subsidiary to the holostratotype.

3. Boundaries

Boundaries of lithostratigraphic units are placed at positions of lithologic change. They are usually designated at sharp lithologic contacts, but also may be placed arbitrarily within zones of lithologic gradation. They should be drawn to express most usefully lithostratigraphic development. Boundaries of lithostratigraphic units commonly cut across time horizons, across the limits of fossil ranges and across the boundaries of any other kind of stratigraphic unit.

Where one rock unit passes vertically or laterally into another by intergrading or complex intertonguing of two or more kinds of rock, the boundary is necessarily arbitrary and should be selected to provide the most practicable assignment of units. For example, in a gradation upward from a limestone unit to a shale unit through interbedded shale and limestone, the boundary may be placed arbitrarily at the top of the highest readily traceable limestone bed in the sequence, or in a gradation laterally from a sandstone unit to a shale unit through increasingly shaly sandstone, the boundary may be placed arbitrarily at the limits of where the rock is still considered to be dominantly a sandstone.

If the zone of intergradation or intertonguing between two units is sufficiently extensive, the rocks of intermediate or mixed lithology may form the basis for establishing and naming a third independent unit, or they may be assigned to an informal provisional unit representing the lateral transition between the two formations and bearing the names of both, separated by a hyphen—Alpha-Beta formation—whichever procedure gives the most meaningful picture of the situation (see Section 3.L, p. 21). In general, the number of units should be kept to a minimum compatible with the greatest practical utility.

Because the many lithologic variations of rock strata offer a wide choice in drawing boundaries of lithostratigraphic units, the selection of these boundaries may be properly influenced by such other factors as lateral traceability, physiographic expression, fossil content, lithogenesis, and well log character, as long as the requirement of a substantial degree of lithologic homogeneity is maintained.

4. Unconformities and Hiatuses

A sequence of rocks of closely similar lithology but including a local or minor hiatus, disconformity, or unconformity should not be separated into more than one lithostratigraphic unit merely because these types of sedimentary breaks occur, unless there is a lithologic distinction adequate to define a boundary. However, the union of adjacent strata separated by regional unconformities or major hiatuses into a single lithostratigraphic unit should preferably be avoided even if no more than minor lithologic differences can be found to justify the separation.

E. PROCEDURES FOR EXTENDING LITHOSTRATIGRAPHIC UNITS—LITHOSTRATIGRAPHIC CORRELATION

A lithostratigraphic unit and its boundaries should be extended away from the type locality only as far as the definitive lithologic features on which the unit was based in its type section are known certainly to exist or through indirect evidence are presumed with assurance to continue.

1. Use of Indirect Evidence for Identification of Units and Their Boundaries

Where lithologic identity is probable but difficult to demonstrate with absolute certainty because exposures are poor or lacking, the unit may be identified and correlated on the basis of criteria which only indirectly indicate lithologic composition. Geomorphic expression, lithogenetic evidence, electric log character, seismic reflections, and distinctive vegetation have been used commonly in this manner. The occurrence of distinctive fossils has also been useful in establishing the probable presence of a lithostratigraphic unit. However, it should always be recognized that tracing or identification of a lithostratigraphic unit by such means is using only inferred evidence of lithology and is not based directly on truly lithostratigraphic criteria.

2. Marker Beds Used as Boundaries

The top or the base of a marker bed may be used as a boundary for a formal lithostratigraphic unit where the marker bed occurs at or near a recognizable vertical change in lithology. Even though the marker beds may be traceable beyond the type area, an extension of the potential boundary markers does not alone justify the extension of a lithostratigraphic unit. Where the rock between two boundary marker beds becomes substantially different from that of the type locality, a new unit should be recognized even though the marker beds continue.

F. NAMING OF LITHOSTRATIGRAPHIC UNITS

The name of a lithostratigraphic unit should be formed from the name of an appropriate local geographic feature, combined with the appropriate term for its rank (group, formation, member, bed), or with the name of the dominant rock type of which the unit is composed, or with both (e.g., Gafsa Formation, Spiti Shale, Sables de Cuise, Taylor Coal Member).

Descriptive adjectives, such as soft, hard, folded, and brecciated, should not be included in the name of a lithostratigraphic unit unless they are part of the lithologic term. The adjectival endings "ian" or "an" should not be used for a lithostratigraphic unit since they are customarily reserved for chronostratigraphic units.

After the complete name of a lithostratigraphic unit has been referred to once in a description or discussion, part of the name subsequently may be omitted to avoid cumbersome repetition, if the omission is compatible with clarity. For example, the Burlington Limestone Formation may be referred to as "the Burlington", "the Limestone", or "the Formation."

Where a lithostratigraphic unit changes to a different lithologic type due to metamorphism or diagenesis, the need for a change in geographic name depends on the degree of change, the persistence of the change, and the assurance of correlation and continuity. For example, the Compton Shale, when regionally metamorphosed to a schist, might better be called the Millikan Schist, not the Compton Schist; but local changes of the Galena Limestone to a dolomite might be referred to more meaningfully as Galena Dolomite than under a new geographic term. As in the case of so many other stratigraphic problems, no hard and fast rule can be given, and decision should be based on common sense and concern for clarity and accuracy.

1. Geographic Component of Name

a. Source. The geographic name should be the name of a natural or artificial feature at or near which the lithostratigraphic unit is typically developed. Names derived from such impermanent sources as farms, churches, schools, crossroads, and small communities are not entirely satisfactory but are acceptable if no others are available. Names for formations or other important rock units may be selected from those that can be found on state, provincial, county, forest service, topographic, hydrographic, or comparable maps. If a name that does not meet this test must be used, the place from which the name is derived should be described and identified precisely and shown on a map accompanying the description of the new unit. A unit should not be named from the source of its components; for example, a till deposit, supposedly derived from the Keewatin glaciation center, should not be called "Keewatin Till." The name of a high-ranking unit may appropriately, though not necessarily, be derived from a more extensive geographic feature or area than the names of its lower ranking components.

b. Duplication.   Duplication of geographic names should be avoided (see Section 3.N, p. 21).

c. Names for Parts of Units.   The same name should not be applied both to the unit as a whole and to a part of it. For example, the Astoria Group should not contain an Astoria Sandstone, nor the Germav Formation a Germav Shale Member.

In formal lithostratigraphic nomenclature the same geographic name should not be used for lithologically different parts or members of a unit. The application of identical geographic names to several minor units in one vertical sequence is considered informal nomenclature (lower Mount Savage coal, Mount Savage fireclay, upper Mount Savage coal, Mount Savage rider coal, and Mount Savage sandstone). The application of identical geographic names to the several lithologic units constituting a cycle of sedimentation is likewise considered informal.

The terms lower, middle, and upper should not be used for formal subdivisions of a lithostratigraphic unit.

d. Spelling.   Spelling of the geographic component of a lithostratigraphic name should conform to the usage of the country that contains the type locality. However, a stratigraphic name repeatedly published with a spelling different from that of its geographic source should nevertheless be retained. For example, Bennett Shale, uniformly used for more than 30 years, should not be altered to Bennet Shale on the grounds that the town is named Bennet. Stratigraphic names that have been spelled variously should be made uniform by adopting the spelling used by the most authoritative local geographical and geological sources. The geographic component of a name should not be altered by translation into an equivalent but different word in another language. For example, Cuchillo should not be translated to Knife, and La Peña should retain the tilde; Canyon should not be translated as Cañon or vice versa, and a lithostratigraphic unit should not be named Montchauve after Bald Mountain. It is proper, however, to translate the lithologic term or rank term; thus the Edwards Limestone may be called Caliza Edwards, and Formación La Casita may be called the La Casita Formation; or Redkinskaya Svita may be called Redkino Formation (but not Redkinskaya Formation).

e. Changes in Geographic Names.   Change in the name of a geographic feature does not entail change of the corresponding name of a stratigraphic unit. The original name of the unit should be maintained; e.g., Mauch Chunk Shale should not be changed to Jim Thorpe Shale because the former town of Mauch Chunk is now called Jim Thorpe.

Disappearance of a geographic feature does not require the elimination of the corresponding name of a stratigraphic unit. For example, Thurman Sandstone, named from a former village in Pittsburgh County, Oklahoma, does not require renaming.

f. Inappropriate Names.   A name that suggests some well-known locality, region, or political division should not, in general, be applied to a unit typically developed in another less well-known locality of the same name. For example, it would not be advisable to use the name "Chicago Formation" for a unit in California, or the "London Formation" for a unit in Wales, even though localities with these names do exist in California and in Wales.

g. Names for Offshore Lithostratigraphic Units.   Offshore drilling in many areas presents problems in applying formal geographic names to the units penetrated in the wells. In some cases, the lithostratigraphic units penetrated in the wells cannot be correlated with the surface units in the bordering region, and it may be very difficult to find locality names for new formations. If the offshore well in which a new lithostratigraphic unit has been penetrated has been given a name taken from coastal, oceanographic, or other features, this name can be used for a subsurface unit providing that the requirements of Section 3.C are followed. However, it may often be necessary to use purely arbitrary nomenclatural designations for offshore subsurface units. The same rules apply to units recognized in underwater mapping.

2. Lithologic Component of Name

Where a lithologic term is used in the name of a lithostratigraphic unit, the simplest generally acceptable term is recommended (e.g., limestone, sandstone, shale, tuff, granite, quartzite, serpentinite). However, a compound term such as "shale-sandstone" or a descriptive adjective, if part of the lithologic term, may be used in special cases where a more detailed indication of composition would be particularly helpful without making the name awkward or misleading in scope. Lithogenetic terms such as "turbidite" or "flysch" should be avoided for formally named lithostratigraphic units.

For intrusive igneous rocks, the lithologic term should express the name of the dominant rock type, for example, Dido Granodiorite. "Dike," "stock," "pluton," "batholith," and other similar names, or more general terms such as "intrusion," are neither lithologic nor stratigraphic terms; accordingly, the names of such intrusive igneous bodies as the Idaho batholith or the Ordubad pluton should not be considered stratigraphic terms.

A series of laterally discontinuous bodies having approximately the same lithologic character, stratigraphic position, and age may be named as a single unit. An example is the Gila Conglomerate, which is a series of valley beds distributed in disconnected areas along the gorges of the upper Gila River. Similarly, a series of disconnected reef limestones or of coal lentils lying at approximately the same stratigraphic position may all be included in the same named unit if their size and separation is not sufficient to warrant naming each one individually. Examples might be the Permian patch-reefs of West Texas or the Leduc reefs of Canada.

3. Preservation of Traditional Names

Although it is strongly urged that all new lithostratigraphic units be named in accordance with the recommendations of this chapter, it is realized that there are many well-established and traditionally used, lithostratigraphic units of long historical standing for which exception should be made. Examples are Millstone Grit, Kupferschiefer, Tea-green Marl, Calcaire Grossier, and such. Such units should not be abandoned merely because they lack geographic names. It is suggested that national stratigraphic bodies make recommendations concerning the conservation of such units under their original names, but it is also recommended that, at the same time, detailed definitions and descriptions be published, and specific stratotypes be designated so that those units that are preserved will have a clear meaning.

G. REVISION OF LITHOSTRATIGRAPHIC UNITS

1. Change in Lithologic Designation

A change in the lithologic term applied to a lithostratigraphic unit does not necessarily require a new geographic term. Priority should not inhibit more exact lithologic designation if the original designation is not everywhere applicable; for example, if the term "Limestone" in such names as Galena Limestone and Bahram Limestone were locally inapplicable, it could be changed to "Dolomite" even though the type section may have been correctly named. If the lithologic variation warrants neither name, the term "Formation" may be preferable.

2. Change in Rank

Change in rank of a lithostratigraphic unit does not require alteration of the geographic part of its name. It is possible for a member to become a formation, or vice versa, and for a formation to become a group or vice versa. For example, the Abbottabad Formation of early work in Pakistan became, in later work, the Abbottabad Group, containing several formations; the Madison Formation of Montana has become the Madison Group; and the Santa Anita Formation of eastern Venezuela has become the Santa Anita Group.

When a previously established formation is broken down into two or more component units that are formally given formation rank, the original formation, with its original geographic name, should be either raised to group status or abandoned. The old name should not be retained for any of the divisions of the original unit.

Examples of change in lithostratigraphic rank from area to area are as follows: the Mila Formation of northern Iran is recognized as a group in eastern Iran; the Osgood Formation, Laurel Limestone, and Waldron Shale of Indiana are classed as members of the Wayne Formation in a part of Tennessee; and the Virgelle Sandstone is a formation in western Montana and a member of the Eagle Sandstone in central Montana.

Chapter Six
Biostratigraphic Units

A. PURPOSE OF BIOSTRATIGRAPHIC CLASSIFICATION

The purpose of biostratigraphic classification is to organize rock strata systematically into named units based on content and distribution of fossils.

The relationship of biostratigraphic units to other kinds of stratigraphic units is discussed in Chapter 8.

B. NATURE OF BIOSTRATIGRAPHIC UNITS

1. Bases for Units

Rock strata are classified biostratigraphically by dividing them into units distinguished by differences in their fossil content. A biostratigraphic unit may be based simply on the presence of fossils as contrasted with their absence; on all kinds of fossils taken together or only on fossils of a certain kind; on the complete assemblage of fossils characterizing a certain stratigraphic interval or only on selected taxons*; on a particular natural association of fossils; on the range of a fossil taxon or fossil taxons; on frequency and abundance of fossil specimens; on certain morphological features of fossils; on fossil indications of mode of life or habitat; on stages of evolutionary development; or on variations in any of the many other features related to the fossil content of strata.

There are thus many different kinds of biostratigraphic units depending on the paleontological feature considered. Biostratigraphic

---

* The plural of taxon is controversial—taxons or taxa. The *Guide* uses *taxons* because the word taxon is an English invention (not a classical Greek word) and the plural ending in *s* is also in keeping with most similar words such as neutrons, ions, morons, icons. However, many prefer to give the word a pseudo-Greek flavor by making the plural into *taxa*, and this procedure is in fact used in the Zoological and Botanical Codes.

units, like lithostratigraphic units, are relatively objective products of classification in that they are based on directly observable features in the rock strata. A *biostratigraphic unit may be considered to be present only within the observed limits of occurrence of the paleontologic feature on which it is based* (see Figure 12, p. 69).

2. Distinctive Nature of Units

Biostratigraphic units are distinct from many other kinds of stratigraphic units in that they are based on discrete particles in the rocks (fossil remains) of almost infinite variety which are disseminated in widely different degrees of density throughout much, but not all, of the Earth's stratigraphic sequence. They are also distinctive in that, as a whole, they show significant evolutionary changes in character with geologic time.

All parts of all rock strata have lithostratigraphic character, and all parts of all rock strata have chronostratigraphic character, but many parts of the Earth's stratigraphic sequence lack significant fossil remains and thus lack biostratigraphic character and are not amenable to biostratigraphic classification.

3. Continuity of Fossils in Relation to Outer Limits of Units

Fossils usually constitute only a minor, disseminated, fractional part of a rock stratum. Even within fossiliferous sequences fossils are rarely found in every bed or formation, nor are they found everywhere along a bed or formation. There are barren spaces or intervals in all sequences, and frequently fossil specimens are widely separated from each other within a fossiliferous bed.

Unless the diagnostic paleontologic elements of a biostratigraphic unit are present, mere similarity of strata in age, or lithology, or environment of deposition to a given biostratigraphic unit does not justify inclusion in that unit. However, if the rock strata lie *within* the outer limits (vertical and lateral) of occurrence of the identifying fossils of the unit, these strata may justifiably be assigned to that unit even though all parts may not show those fossils. The extent to which continuity of occurrence of zone-index fossils may be interrupted, without the intervening strata being excluded from the zone, is a subjective matter for which it would be difficult to lay down strict rules.

4. Significance of Fossils

Because of their complexity, variety, and local abundance, fossils are often important simply as distinctive lithologic features of rock strata.

As the remains of once living forms, they are, moreover, sensitive indicators of past environments of deposition. Finally, because of the progressive, nonrepetitive, and more or less orderly evolution of life forms, fossils are particularly valuable in time-correlation of strata and in placing strata in their proper position with reference to the world geochronological scale.

5. Life Communities and Death Assemblages

The fossils found in sedimentary strata are either remains of organisms that lived in the area and were covered by its deposits; or remains of organisms that were brought together in the area only after death, by currents, by settling in a water body, or by other means. They are commonly a mixture of the two categories. Either or both categories of fossils may be the basis for biostratigraphic zonation.

6. Reworked Fossils

Fossils from rocks of one age frequently have been eroded, transported, and redeposited in sediments of younger age. The reworked fossils may thus be mingled with indigenous fossils, or they may constitute the only fossils appearing in the new sediment. In some cases, the reworked fossils can be distinguished readily from the indigenous ones, but in other cases they cannot. This last is particularly true in the case of micro- or nannofossils, where a fossil specimen may behave like a single grain of sediment and pass through one or more cycles of sedimentation with little evidence of wear.

All fossil remains, whether indigenous or reworked, may constitute distinctive features of a sediment and may serve as the bases for biostratigraphic zonation; however, because of the difference in their significance with respect to age and environment, fossils that can be identified as reworked should be treated apart from those believed to be indigenous.

7. Introduced or Infiltrated Fossils

Under some circumstances, rocks may contain fossils younger than the enclosing material. Sometimes this is due to infiltration of fluids carrying micro- or nannofossils from one formation into the pore spaces or fractures of an underlying formation. It also happens that animal burrows or root cavities extending down into one formation may be filled with fossiliferous material from an overlying formation. Likewise, sedimentary dikes or diapirs may contaminate a formation with either younger or older fossil material. Such introduced fossils should be distinguished from indigenous fossils in biostratigraphic zonation.

8. Overlaps and Gaps Between Units

There are frequently overlaps or gaps, both vertically and horizontally, among different kinds of biostratigraphic units, as well as among similar kinds of units based on different taxons or different kinds of fossils.

9. Effects of Stratigraphic Condensation

Extremely low rates of sedimentation may result in fossil representatives of different ages and different environments being mingled or very intimately associated in a very thin stratigraphic interval, even in a single bed.

C. DEFINITIONS

1. Biostratigraphy—the element of stratigraphy that deals with the remains or evidences of former life in strata and with the organization of strata into units based on their fossil content.

2. Biostratigraphic Classification—the organization of strata into units based on their fossil content.

3. Biostratigraphic Unit—a body of rock strata unified by its fossil content or paleontological character and thus differentiated from adjacent strata. A biostratigraphic unit is present only within the limits of observed occurrence of the particular biostratigraphic feature on which it is based.

4. Biostratigraphic Zone (Biozone)—a general term for any kind of biostratigraphic unit. *Biozone* is a short alternative term for biostratigraphic zone. *Bio* should be used in front of the term zone to distinguish biostratigraphic zones from other kinds of zones whenever there is any danger of confusion. This is particularly important in preventing confusion between biozones and chronozones; both may be named from a fossil or fossils, but they are quite different from each other in concept (Figure 12, p. 69).

Biozones vary greatly in thickness and geographic extent. They may range from a local bed to a unit thousands of meters thick or extending worldwide. The total time represented by a biozone may be referred to simply as its time or time-value, or its *biochron*.

5. Superzones and Subzones. Several biozones with common biostratigraphic features may be grouped into *superzones* (*superbiozones*); any kind of biozone may be divided into *subzones* (*subbiozones*) to express finer biostratigraphic detail. A zone need not be completely subdivided into subzones. Subzones have also been divided into very small units called *zonules*.

6. Barren Interzones and Intrazones. Intervals lacking in fossils between successive biozones may be called *barren interzones* and referred to informally by reference to the adjacent biozones, for example, *Exus parvus* to *Exus magnus* barren interzone. Similarly, barren intervals of substantial thickness *within* biozones may be referred to as *barren intrazones*, for example, the barren intrazone near the top of the *Exus albus* Assemblage-zone.

7. Biohorizons—*surfaces* of biostratigraphic change or of distinctive biostratigraphic character; preeminently valuable for correlation (not necessarily time-correlation); commonly used as a biozone boundary, though often recognized as horizons *within* biozones. In theory, a biohorizon is strictly a surface or interface; in practice, the term may refer to a thin but biostratigraphically distinctive bed. Features on which biohorizons are commonly based include "first appearances," "last occurrences," distinctive occurrences, changes in frequency and abundance, evolutionary changes, and changes in the character of individual taxons (e.g., changes in direction of coiling in foraminifers, or in number of septa in corals).

Biohorizons have been called surfaces, horizons, levels, limits, boundaries, bands, markers, indexes, datums, datum planes, datum levels, key horizons, key beds, marker beds, and so on. This Guide favors the term biohorizon because it has counterparts in the terms lithohorizons and chronohorizon, and it usually requires no translation from one language to another.

D. KINDS OF BIOSTRATIGRAPHIC UNITS

1. General

Biozone is the general term for any kind of biostratigraphic unit. However, since strata can be zoned biostratigraphically in many different ways, there are different kinds of biozones, each having a different significance and each being useful under appropriate circumstances. It is important, therefore, to have separate, specific, and well-defined terms for each in order to indicate clearly which kind of biozone is being used.

The following four general types of biozones are in common use:

a. Assemblage-Zone—a group of strata characterized by a distinctive *natural assemblage* of all forms present or of the forms present of a certain kind or kinds.

b. Range-Zone—a group of strata representing the *stratigraphic range of some selected element* of the total assemblage of fossil forms present.

c. Acme-Zone—a group of strata based on the *abundance* or development of certain forms, regardless of either association or range.

d. Interval-Zone (Interbiohorizon Zone)—the stratigraphic *interval* between two biohorizons.

"Assemblage," "range," and "abundance" are, of course, not mutually exclusive criteria, and a single stratigraphic section can be subdivided independently into assemblage-zones, range-zones, and acme-zones, depending on the feature to be emphasized.

2. Assemblage-Zones (Cenozones)

See Figure 4.

a. Definition and Significance. A biostratigraphic *assemblage-zone* is a body of strata whose content of fossils, or of fossils of a certain kind, taken in its entirety, constitutes a *natural* assemblage or association that distinguishes it in biostratigraphic character from adjacent strata.

(The assemblage-zone has also been known as a *cenozone*, from the Greek *koinos* meaning common. Although this term has the advantage of being derived from a classical language, it does not satisfactorily express the significance of an assemblage-zone and for this reason the English term seems preferable.)

An assemblage-zone may be based on all kinds of fossil forms present, or it may be restricted to forms of only certain kinds. Thus we may have an assemblage-zone based only on fossil *fauna* or one based only on fossil *flora*; an assemblage-zone of corals, or of foraminifers, or of mollusks, or of dasyclad algae; an assemblage-zone of planktic forms or an assemblage-zone of benthic forms; and so on. The basis on which the zone is founded should be made clear either in the name of the assemblage-zone or in an accompanying explanation. Whatever its basis, an assemblage-zone is characterized by an assemblage or association of fossil forms (of one specified fossil group, of many groups, or of all groups combined) that are assumed to have lived together or to have

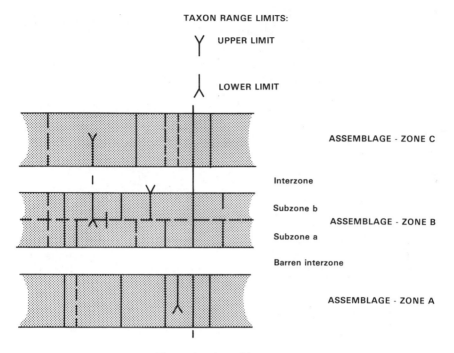

Figure 4.   Assemblage-zones

died together or to have accumulated together, or, in any case, to have been entombed together.

Many factors complicate the interpretation of assemblages. Although the entirety of fossil remains found in any body of rock strata must be looked on as a natural and objective feature of these strata, the possibly diverse origins of the assemblage should be fully recognized. Some part may represent the remains of indigenous bottom life (but much of this life left no fossil record); some part may have settled to the bottom from overlying waters (but some part may also have been lost by solution as it settled); some part may have been brought in by currents after death (but some part may also have been carried away by currents); some part may be the result of mixing of the constituents of different facies. These factors indicate the need for analysis and discrimination in arriving at the most useful assemblage-zones. For example, the same interval of strata might be zoned quite differently on its planktic elements than on its benthic elements. A total general assemblage may thus be broken down profitably into many specifically significant component assemblages.

The scope and character of an assemblage-zone may be defined in words by giving the names of all or many of the principal taxons of which it is composed. However, its concept is best transmitted by additionally designating a stratotype, which serves as protection against the inadequacies of language and the incompleteness of fossil collection, and which can be used as a standard of reference in the identification of the assemblage-zone elsewhere.

Assemblage zones are usually linked in practice to local areas or regions, as they are closely linked with life environments that vary greatly geographically. However, marine planktic fossil assemblages may approach worldwide extent within restricted latitude ranges and under conditions where variation in temperature is low.

Assemblages may recur many times in a stratigraphic sequence with very little overall change, representing recurrences of essentially the same environment in a short period of time. On the other hand, over any considerable period of geologic time, evolutionary changes have been sufficient to make the assemblages of one age distinctive from those of another.

The assemblage-zone is particularly significant as an indicator of environment. It is also a general indicator of geologic age, although it is not controlled by total stratigraphic range of taxons. Regardless of interpretation, it is an observable, relatively objective biostratigraphic unit of high value in local correlation.

b. Boundaries.   Assemblage-zone boundaries are drawn at surfaces (biohorizons) marking the limits of occurrence of the assemblage characteristic of the unit. The accuracy with which the boundaries of an assemblage-zone may be identified obviously depends largely on the precision of the assemblage definition. Not all members of the association need occur before strata can be assigned to the zone. Identification of the zone and of its boundaries is thus a matter of interpretation and judgment. The total range of any constituent taxon may extend beyond the boundaries of the assemblage-zone.

c. Name.   The name of an assemblage-zone should preferably be derived from two or more of the prominent and diagnostic constituents of the fossil assemblage, for example, *Eponides-Planorbulinella* Assemblage-zone. Not all parts of a fossiliferous stratigraphic sequence need have named assemblage-zones. Recurrent assemblage-zones or subtle differences within one assemblage-zone can be differentiated by words like lower, middle and upper, or by letters or numbers.

3. Range-Zones

A biostratigraphic *range-zone* is the body of strata representing the total *range* of occurrences of any *selected* element of the total assemblage of fossil forms in a stratigraphic sequence. The word "range" implies extent in both horizontal and vertical directions.

(The word "range-zone" is difficult to translate into some languages. *Acrozone,* from the Greek *akros,* meaning topmost or extreme, has been suggested as a substitute for range-zone in order to have a term derived directly from a classical language. However, the term is not very informative as to the nature of the zone and might even be misleading. For this reason this *Guide* recommends the use of the English term in all languages.)

A biostratigraphic range-zone may represent the stratigraphic range of some one taxonomic unit (species, genus, family, order, etc.), or of a grouping of taxons, or of a lineage or segment of a lineage, or of any particular paleontological feature whatsoever. Its meaning should be made clear either in the name of the range-zone or in an accompanying explanation. While the accuracy of identification and biologic description of the taxons on which the zone is based is critical to its value, there is always a degree of subjectivity and impermanence involved in taxonomic identification. Also a considerable variation in the range of a taxon may depend on whether its limits are defined morphotypically or by statistical population studies.

There are many kinds of range-zones. Some of the principal kinds are discussed below.

a. Taxon-Range-Zone. See Figure 5.

i. Definition and Significance. A *taxon-range-zone* is the body of strata representing the total range of occurrence (horizontal and vertical) of specimens of a particular taxon (species, genus, family, etc.). Thus the Range-zone of *Linoproductus cora* is the total body of strata enclosed

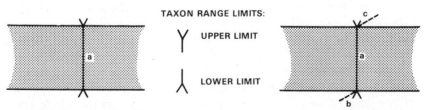

Figure 5. Taxon-range-zones. (*a*) Range of taxon. (*b, c*) Ancestral and descendant taxons.

within the outer limits of the established occurrence of specimens of *Linoproductus cora,* and the Range-zone of *Globotruncana* is the total body of strata enclosed within the outer limits of the established occurrence of specimens of any species whatsoever of *Globotruncana.* The range-zone of a taxon means its maximal geographic and stratigraphic extent, unless a more restricted area is specifically indicated, for example, the Range-zone of *Linoproductus cora* in Europe or the Range-zone of *Exus albus* in the Quai River section. The terms *genus-zone* and *species-zone* have occasionally been used in the same sense as taxon-range-zone.

The taxon-range-zone is particularly valuable as an indicator of geologic age because of the limited time range (life spans) of specific taxons through geologic history. Thus the assignment of a stratum to the range-zone of graptolites or to the Range-zone of *Lepidocyclina pustulosa* places it in a certain specific position in the course of Earth history. The range-zone of a taxon may also be significant of environment, insofar as the distribution of specimens of that taxon was controlled by environment.

The taxon-range-zone is as objective as the concept of the taxon on which it is based, and it may be as global in extent as the occurrence of the individual taxon. Both the geographic and the stratigraphic extent of a taxon-range-zone tend to increase with ascending rank in the taxonomic hierarchy of the taxon selected.

Taxon-range-zones do not lend themselves to a complete and systematic partitioning of all strata into a single set of units without gaps or overlaps because of the numerous gaps and overlaps in the ranges of the great number of taxons fossilized in rock strata.

ii. Boundaries. The boundaries of a taxon-range-zone are surfaces (biohorizons) marking the outermost limits of known occurrence anywhere of specimens of the taxon whose range is to be represented by the zone. Thus the limits are those of the origin and extinction of the taxon insofar as these are known. However, in any one section the boundaries are simply the horizons of first appearance and final disappearance in that section. Both of these may be facies-controlled or hiatus-controlled in that particular section; hence the true overall limits of the range-zone will not be found until all local sections have been examined. The very abrupt appearance or disappearance of a form in a sequence of strata often indicates local facies control over its occurrence or the presence of a hiatus; only when a section shows a gradation from the immediate ancestral forms of the taxon below to the immediate descendants above is there any assurance that the whole vertical range in an area is represented.

The boundaries of a taxon-range-zone are continually subject to change with new discoveries. Moreover, they may not be truly representative of the original taxon range due to loss of specimens by subsequent solution or metamorphism. Also, the boundaries of a named taxon-range-zone are subject to change with changes in the scope of the taxon.

iii. Local Range of a Taxon.   The terms *teil-zone, local-range-zone,* and *topozone* have been used to indicate the range of a taxon *in some particular area* as contrasted with total range. However, the range in a local area is not meaningful unless the name of that area is given. The *Guide,* therefore, suggests that these terms not be used. Instead, reference should be made, for instance, simply to the range-zone of taxon-a in the Mediterranean region, or in the Grand Canyon section, without any additional modifying zonal term.

iv. Name and Reference Sections.   The taxon-range-zone is named from the taxon whose range it expresses, for example, *Didymograptus* Range-zone, range-zone of mammals, *Globigerina brevis* Range-zone. There is no need of a stratotype for a taxon-range-zone; the concept of the zone rests entirely on the concept of the taxon and is independent of any specific section of strata. However, reference sections to demonstrate the actual occurrence of the taxon as claimed are useful.

b. Concurrent-Range-Zone.   See Figure 6.

i. Definition and Significance.   A *concurrent-range-zone* is defined as the concurrent or coincident parts of the range-zones of two or more specified taxons selected from among the total forms contained in a sequence of strata. Definition of a concurrent-range-zone does not require that the range-zones of all taxons present concur or overlap, nor does it require that all those that do overlap be considered. The

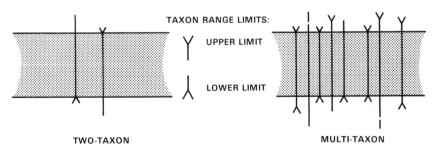

Figure 6.   Concurrent-range-zones

objective in defining the concurrent-range-zone is to select those over-lapping range-zones that represent a biostratigraphic unit with op-timum time significance and possibilities of geographic extension, al-though the zone is not in itself a chronostratigraphic unit.

Strictly speaking, all the taxons named in the definition of a concurrent-range-zone should be present before the zone can be rec-ognized legitimately. In common practice, considerable tolerance is allowed, and the zone is recognized on the basis of the approximately joint occurrence of only a substantial proportion of the specified index taxons (cf. Oppel-zone, Section 6.D.3, p. 57).

The principle of the concurrent-range-zone has long been used in time-correlation. The use of two or more taxons whose range-zones overlap reinforces the time significance of an individual taxon-range.

The term concurrent-range-zone is expressive of the meaning of the zone, although difficult to translate from English into some other languages. Concurrent-range-zones have been referred to as *overlap zones* or *range-overlap zones*.

ii. Boundaries.   The boundary of a concurrent-range-zone is the outer limit of the concurrence of the selected taxons designated as diagnostic of the zone. Drawing the boundaries of concurrent-range-zones pre-sents problems that require thorough study of range charts and geo-graphic distribution of the taxons, and very carefully considered selec-tion and rejection of taxonomic elements. If only two taxons are considered diagnostic of the zone, the matter of boundaries is relatively simple. However, with *more* than two taxons named as diagnostic of the zone, the matter of boundaries becomes increasingly complex. Figure 7 shows that if a concurrence of all five taxons is necessary for recogni-tion of the zone, the extent of the concurrent range might be so much reduced both vertically and horizontally over that of any of the in-dividual taxons as to make its time span and geographic extent ex-tremely small. On the other hand, if the number of taxons required to be present for identification of the zone is limited selectively to fewer than the five taxons designated as diagnostic, the problem arises as to how many and which ones should be chosen. The figure shows that the boundary of the hypothetical concurrent-range-zone could vary mark-edly depending upon what degree of concurrence of which taxons was required.

iii. Name and Reference Sections.   A concurrent-range-zone is named from two or more of the taxons which characterize the zone by their concurrence, for example, *Globigerina sellii-Pseudohastigerina barbadoensis* Concurrent-range-zone. Although the concurrent-range-zone cannot

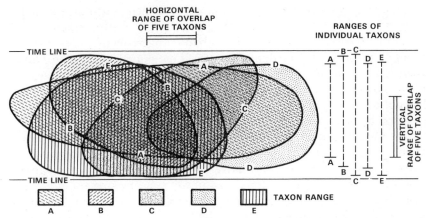

Figure 7. Variations in extent of a concurrent-range-zone depending on number of taxons and degree of concurrence required. Interval between horizontal time-lines represents cross section through a sequence of essentially horizontal sediments. For simplicity, picture is two dimensional, and continuous distribution of each taxon within its pattern (A, B, C, D, or E) is assumed. Vertical lines at right show vertical range for each taxon. Double line at extreme right shows extent of overlap in vertical range for all five taxons. Double line at top shows extent of overlap in horizontal range for all five taxons.

be defined satisfactorily by a stratotype, citation of the reference localities where the unit exists and where the selected taxons are adequately represented is often necessary as field evidence of their concurrent presence.

c. Oppel-Zone. See Figure 8.

i. Definition and Significance. The Oppel-zone, named after the German stratigrapher, Albert Oppel (1831–1865), whose usage it roughly

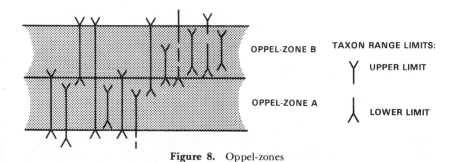

Figure 8. Oppel-zones

follows, largely embodies the concept of the concurrent-range-zone but relaxes its strict interpretation sufficiently to allow supplementary use of biostratigraphic criteria other than range-concurrence that are thought to be useful in demonstrating time-equivalence. The Oppel-zone is a more subjective, more loosely defined, and more easily applied biozone than the concurrent-range-zone. The term Oppel-zone has not been widely used, but the concept corresponds to a widespread practice in biostratigraphic zonation.

The Oppel-zone may be defined as a zone characterized by an association or aggregation of *selected* taxons of restricted and largely concurrent range, chosen as indicative of approximate contemporaneity. Not all of the taxons considered diagnostic need be present at any one place for the zone to be legitimately identified. The lower part of the zone is commonly marked largely by first appearance and its upper part by last appearance of certain taxons. The body of the zone is marked largely by concurrences of the diagnostic taxons.

The Oppel-zone is difficult to define empirically because judgment may vary as to how many and which of the selected diagnostic taxons need be present to identify the zone.

ii. Boundaries. The boundaries of an Oppel-zone are the limits of distribution of the ensemble of fossil forms considered distinctive of the zone. Because of the complexity and indefiniteness of Oppel-zone criteria, boundary positions are to a considerable extent subject to the worker's judgment. Boundaries of adjacent Oppel-zones must often be placed within transition intervals, and different workers might well choose different positions. Sharp boundaries marking simultaneous appearance or disappearance of many of the diagnostic taxons may suggest facies changes, paleogeographic changes, or hiatuses.

iii. Name and Reference Sections. The name of an Oppel-zone is derived from some one prominent taxon, which may or may not be everywhere present in the zone, for example, *Siphogenerinoides bramletti* Oppel-zone. In principle, Oppel-zones should be amenable to subdivision into subzones, or to grouping into larger units (superzones). The Oppel-zone is usually restricted in application to a single biogeographic province. Although the scope of the Oppel-zone cannot be defined by a stratotype, the designation of a reference section is useful.

d. Lineage-Zone (Phylozone). See Figure 9.

i. Definition and Significance. A lineage-zone is a type of range-zone consisting of the body of strata containing specimens representing a segment of an evolutionary or developmental line or trend, defined

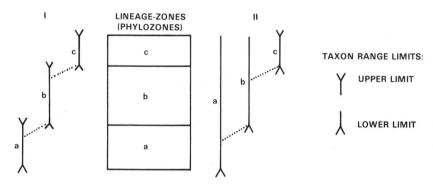

Figure 9. Theoretical examples of lineage-zones or phylozones. In Example I, a, b, and c are range-zones of taxons a, b, and c (or forms a, b, and c within a single taxon). In Example II, a and b are parts of range-zones and c is a complete range-zone. Other examples of lineage-zones could be drawn for different evolutionary patterns.

above and below by changes in features of the line or trend. This type of zone has also been termed an *evolutionary zone,* a *morphogenetic zone,* a *phylogenetic zone,* or a *phylozone.*

The scope of a lineage-zone varies. It depends somewhat on the nature and magnitude of the form-change recognized. Thus it may include a number of successive taxons in evolutionary descent, or it may cover a graded sequence of forms in a single taxon, from the first emergence to descendant transition or extinction. It may correspond to a taxon-range-zone, to a part of a taxon-range-zone, or to a concurrent-range-zone.

In theory, a system of overlapping zones based on several lineages offers one of the best assurances of reliable time-correlation on a biostratigraphic basis. In practice, the assurance may be lessened by uncertainty on evolutionary courses and by the subjectivity of taxonomic identification and morphologic differentiation. Moreover, recognition must be given to variations in rates of evolution and dispersal.

ii. Name. A lineage-zone may most simply be named after the key taxon, whether a transitional form (as I*b* of Figure 9) or the latest form in a slowly evolving lineage (as II*c* of Figure 9). Examples of the two cases are, respectively, the *Globorotalia fohsi fohsi* Lineage-zone and the *Globorotalia cerroazulensis cunialensis* Lineage-zone.

4. Acme-Zones. See Figure 10.

An *acme-zone* is a body of strata representing the acme or maximum development—usually maximum abundance or frequency of

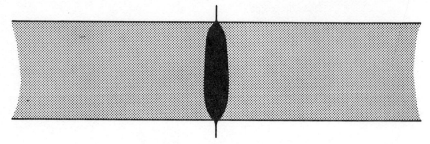

Figure 10. Acme-zone

occurrence—of some species, genus, or other taxon, but not its total range.

A problem is presented in defining what constitutes a "maximum of development," and thus in defining the boundaries of an acme-zone. For instance, "maximum development" may refer to abundance of specimens of a taxon or it may refer to number of species in a genus. Some workers consider acme-zones valuable indicators of chronostratigraphic position.

The acme-zone has been called *epibole* (although epibole in its original definition has no reference to acme), *peak-zone,* and *flood-zone.*

The acme-zone takes its name from the taxon whose zone of maximum development it delimits, for example, *Didymograptus* Acme-zone.

5. Interval-Zones. See Figure 11.

A *biostratigraphic interval-zone* (more properly termed a *biointerval-zone* or an *interbiohorizon-zone*) is an interval between two distinctive biostratigraphic horizons. Such a zone is not itself the range-zone of any taxon or concurrence of taxons; and it may contain no particularly distinctive biostratigraphic assemblage or feature. It may have no more overall significance than that of position between two identifiable biohorizons. The base of such a zone might be marked by the horizon of first appearance of taxon-a or the last appearance of taxon-b, and the top might be marked by the horizon of first appearance of taxon-c or the last appearance of taxon-d. Or the base might be marked by the top boundary of any distinctive biozone and the top by the base of the next succeeding distinctive biozone boundary. The interval-zone is in common use for correlation purposes.

The names given to interval-zones may be derived from the names of the boundary horizons, the name for the basal boundary preceding

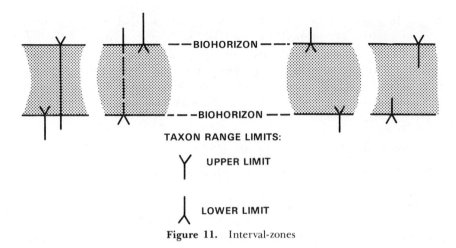

Figure 11. Interval-zones

that for the upper boundary, for example, *Globigerinoides sicanus/ Orbulina suturalis* Interval-zone. However, a name of the type given fails to convey whether the named taxons appear or disappear at the zone limits, or if some other criterion is involved (e.g., abundance, dwarfing or gigantism, coiling direction). Consequently, one may alternatively select the name of a taxon typical of the zone, *though not confined to it,* and use it as a handy label. A representative example is the *Globigerina ciperoensis* Zone which, by definition, is the interval-zone between the extinction-level of *Globorotalia opima opima* and the advent-level of *Globorotalia kugleri,* and occupies only a part of the range of the selected name-fossil.

6. Other Kinds of Biozones

Additional kinds of biozones not mentioned above are defined in the glossary of biostratigraphic terms accompanying ISSC Report 5 (1971), *A Preliminary Report on Biostratigraphic Units.* The number of biozones in use may expand even further as special circumstances call for new terminology. For example, range-zones based on tracks or trails, reworked fossils, or morphologic changes in community structures (as in the case of stromatolites) may not appropriately fall under existing terminology. Still, these units may be of great value.

7. Informal Biozones

The distinction between formal and informal biostratigraphic terms and usage is somewhat vague. In general, a formal biostratigraphic unit is one that is defined and named in accordance with the rules of a conventionally established system of scientific classification and nomenclature. An informal biostratigraphic unit is one that is used in a broad or free sense, without precise connotation and without being part of an organized system of terminology. An example of the latter would be such expressions in the literature as "the beds with oysters" or "marnes à *Cerithium*."

8. Hierarchies in Biostratigraphic Units

The different kinds of biozones described above do not represent different ranks of a biostratigraphic hierarchy—range-zones are not subdivided into assemblage-zones, and so on. Some kinds of biozones, however, may be usefully subdivided into subzones and/or grouped into superzones. Such is the case with assemblage-zones and Oppel-zones, for instance.

With respect to taxon-range-zones, there is no need for a hierarchy of zone terms because the hierarchical system of biological taxonomy extends also to these biostratigraphic units in the sense that the range-zone of a species is subsidiary to the range-zone of the genus to which it belongs, and so on.

E. PROCEDURES FOR ESTABLISHING BIOSTRATIGRAPHIC UNITS

General procedure for establishing stratigraphic units is discussed in Section 3.B, p. 16-17, and the procedure for biostratigraphic units accords closely with that for other stratigraphic units in most respects. Special mention is made again, however, of the need to specify the *kind* of biozone being proposed and the basis used for defining its limits. Also, it is emphasized that figures and descriptions of the taxons diagnostic of a unit should be provided or references to the literature in which they can be found should be given.

In setting up new zones or in selecting for use zones that already have been proposed, practicability in correlation should be considered. Other things being equal, units based on abundant, widespread, and readily recognized taxons should be preferred. Highly esoteric zonal criteria lose much of their value because of difficulty of application.

Stratotypes are useful in defining some, but not all, biostratigraphic units. Those units whose meaning can be expressed by reference to the fossil content of a specific and limited section of strata can be better defined by reference to stratotypes as well as by verbal descriptions. For example, the concept of a certain biostratigraphic assemblage-zone is generally derived from the real occurrence of a certain assemblage in a certain sequence of strata. The designation of some one specific and delimited section in this sequence as the type expression of this assemblage provides a useful aid to the definition and understanding of this assemblage-zone and its recognition elsewhere.

On the other hand, biostratigraphic units whose scope and extent are based on a concept independent of any specific section of strata cannot be referred satisfactorily to any specific stratotype for definition. A biostratigraphic range-zone and some concurrent range-zones are examples of units of this sort. The concept of a range-zone is the total body of strata encompassed by the range (horizontal and vertical) of specimens of a certain taxon, and this concept cannot be controlled or defined by any designated section of strata. Instead, it is the concept of the taxon and its occurrence, rather than any specific stratigraphic section, that provides the control for the unit.

F.   PROCEDURES FOR EXTENDING BIOSTRATIGRAPHIC UNITS—BIOSTRATIGRAPHIC CORRELATION

Biostratigraphic units are extended away from their type localities by biostratigraphic correlation, which is the establishment of correspondence in character and in biostratigraphic position between geographically separated units or horizons (marker beds) based on their fossil content. Biostratigraphic correlation is not necessarily time-correlation. It may coincide with time-correlation, or it may be a facies correlation and therefore diachronous. In either case, correlation is a matter of judgment. No two separated stratigraphic sequences have points that are identical in fossil content, and a subjective assessment on how closely two separated fossil occurrences must correspond, both in content and in stratigraphic position, is always required before they can justifiably be said to correlate with each other.

G.   NAMING OF BIOSTRATIGRAPHIC UNITS

The formal name of a biostratigraphic unit should be formed from the names of one or more appropriate fossils combined with the appropriate term for the kind of unit in question, modified, if necessary, to show its rank, for example, *Exus albus* Assemblage-zone. (For correct printing of fossil names see Section 3.R,4 p. 23). The same name should not be used for different biostratigraphic units of the same kind, even if of different rank.

A disadvantage of names formed from more than one taxon is their often cumbersome length (e.g., *Globorotalia (Turborotalia) acostaensis acostaensis—Globorotalia (G.) merotumida* Partial-range-zone). This difficulty has commonly been circumvented by naming the zone for a single taxon that is common in the interval although not otherwise diagnostic of the zonal limits. For example, the *Praeglobotruncana gigantea* (Interval-) Zone includes only that part of the name-species range between extinction of *Rotalipora cushmani* and advent of *Globotruncana helvetica.* Single taxon names of the type indicated may be considered formal, provided their introduction is accompanied by an unequivocal designation of the zonal limits.

Codification of biostratigraphic zones by letters or numbers or a combination of both is becoming a common practice. Code designations are brief, and they avoid repetition of the lengthy formal names of zones (an advantage in both written and verbal presentations); also, letter and/or number sequences automatically indicate the sequence and relative positions of the zones (not true of their formal names); and finally, code-designations facilitate liaison between geologists and other professionals such as engineers.

On the other hand, code-designations are inflexible and do not lend themselves readily to insertions, combinations, or eliminations within a zonal sequence once it has been published; letter/number designations have no intrinsic meaning, and confusion can arise if two or more stratigraphers have applied them in different senses within the same general area.

In any case, code-designation of biostratigraphic units should be considered informal.

H. REVISION OF BIOSTRATIGRAPHIC UNITS

Revision of stratigraphic units in general is discussed in Section 3.G, p. 20, and the basic rules of priority are discussed in Section 3.E, p. 19. In the case of biostratigraphic units, it must be kept in mind that out of the almost limitless number and variety of overlapping biozones that could be proposed, not necessarily the first to be described and named but the most useful should be preserved. This means that workers must continually be free to propose new zones or improve previous proposals in both scope and nomenclature. In any new proposal or any revision, pertinent previous work should, of course, be recognized. But to try to impose a rigid system of priority would not only be impracticable but would run serious risks of defeating progress. Among critical considerations affecting the adoption of any newly proposed biozone or revision of already existing biozones should be adequacy of description, freedom from ambiguity, and extent of applicability.

Names of biostratigraphic units should preferably be changed to conform with changes in names of taxons required by the *International Code of Zoological Nomenclature* and by the *International Code of Botanical Nomenclature.* Also, named biostratigraphic units should be revised to accord with changes in the scope of taxons that may have occurred subsequent to the naming of the unit. However, a fossil name once used for a zone should normally not be available for use in a different zonal sense by a later author. If a taxonomic term is no longer valid, it should be in quotation marks, for example, *"Rotalia" beccarii* Zone.

Chapter Seven
Chronostratigraphic Units

A. PURPOSE OF CHRONOSTRATIGRAPHIC CLASSIFICATION

The general purpose of chronostratigraphic classification is to organize systematically the Earth's sequence of rock strata into named units (chronostratigraphic units), corresponding to intervals of geologic time (geochronologic units), to serve as a basis for time-correlation and a reference system for recording events of geologic history. Specific objectives are as follows.

1. To determine local time relations. Local time-correlation of strata and the simple determination of the relative age of strata in local sections or areas are important contributions to local or regional geology, regardless of any scheme of organization of strata into named chronostratigraphic units of worldwide application.

2. To establish a Standard Global Chronostratigraphic Scale. A major goal is to establish a complete and systematically arranged hierarchy of named and defined chronostratigraphic units, of both regional and worldwide application. Such a hierarchy would serve as a standard framework for expressing the age of all rock strata and for relating all rock strata to Earth history. Ideally, the named units comprising each rank in this Standard Global Chronostratigraphic Scale should, as a whole, encompass the entire stratigraphic sequence, without gaps and without overlaps.

The relations of chronostratigraphic units to other kinds of stratigraphic units are discussed in Chapter 8.

B. DEFINITIONS

1. **Chronostratigraphy**—the element of stratigraphy that deals with the *age* of strata and their *time* relations.

2. **Chronostratigraphic Classification**—the organization of rock strata into units on the basis of their age or time of origin.

3.   Chronostratigraphic Unit—a body of rock strata that is unified by being the rocks formed during a specific interval of geologic time. Such a unit represents all rocks formed during a certain time span of Earth history and only those rocks formed during that time span. Chronostratigraphic units are bounded by isochronous surfaces. The rank and relative magnitude of the units in the chronostratigraphic hierarchy are a function of the length of the time interval that their rocks subtend, rather than of their physical thickness.

4.   Chronozone—a zonal unit embracing all rocks formed anywhere during the time range of some geologic feature or some specified interval of rock strata. The basis for the time span of a chronozone may be the time range of a biostratigraphic unit, or of a lithostratigraphic unit, or of any other feature of strata that has a time range; or it may be any purely arbitrary but specified interval of strata, provided it has features allowing time-correlation with stratal sequences elsewhere. Chronozones may be of widely different time spans. Thus we may speak of the chronozone of the ammonites, which would include all strata formed in the long interval during which the ammonites existed, regardless of whether the strata contained ammonites; or we may speak of the chronozone of *Exus albus*, a species of very limited time range; or we may speak of the chronozone of the São Tomé volcanic rocks, a unit of very local development but representing a relatively long interval of Tertiary time, which would include all strata formed anywhere during this interval of time whether or not the strata were volcanic rocks.

The chronozone based on the range of a certain taxon should be clearly distinguished from the biozone based on the range of that same taxon (taxon-range-zone). The loose use of the unqualified word "zone" for both has been the source of much confusion. The difference between the concepts of biozone and chronozone is illustrated in Figure 12, p. 69. The *Exus albus* Biozone (a range-zone) is limited in extent to the strata in which specimens of *Exus albus* occur. The *Exus albus* Chronozone (a chronostratigraphic unit) includes all strata, everywhere, of the same age as that represented by the total vertical range of *Exus albus*, regardless of whether specimens of *Exus albus* are present.

The term chronozone may be used formally as a chronostratigraphic unit of minor rank (see Section 7.C.2, p. 69) and as an informal unit of unspecified rank (see Section 7.C.9, p. 75-76).

5.   Chronostratigraphic Horizon (Chronohorizon)—a stratigraphic surface or interface that is isochronous—everywhere of the same age. Although a chronohorizon is theoretically without thickness, the term chronohorizon is commonly applied also to very thin and distinctive intervals that are essentially isochronous over their whole geographic extent and thus constitute excellent time-reference or time-correlation horizons. Chronohorizons have also been called markers, marker horizons, marker beds, key beds, datums, levels, time-surfaces, and so on. Examples of horizons that may have strong chronostratigraphic significance include many biohorizons, bentonite beds resulting from volcanic ash falls, tonsteins, phosphorite layers, horizons of magnetic reversal, coal beds, some electric log markers, seismic reflectors, etc. The geochronologic equivalent of a chronohorizon is a *moment* (or an *instant*, if it has no appreciable time duration).

C. KINDS OF CHRONOSTRATIGRAPHIC UNITS

1. Hierarchy of Chronostratigraphic and Geochronologic Unit-Terms

The *Guide* recommends the following formal chronostratigraphic terms and geochronologic equivalents to express units of different rank or time scope (Table 2).

Table 2. Conventional Hierarchy of Chronostratigraphic and Geochronologic Terms

Chronostratigraphic	Geochronologic
Eonothem	Eon
Erathem	Era
System[a]	Period[a]
Series[a]	Epoch[a]
Stage[a]	Age[a]
Chronozone[b]	Chron[b]

[a] If additional ranks are needed, the prefixes "sub" and "super" may be applied to these terms.

[b] The term chronozone is shown as a formal chronostratigraphic unit of lower rank than a stage but not necessarily an aliquot part of a stage (Sections 7.C.2, p. 69-70 and 7.C.3, p. 71). For a broader informal use of this term see Section 7.C.9, p. 75-76.

Position *within* a chronostratigraphic unit is commonly best indicated by such adjectives as: basal, lower, middle, upper, uppermost; whereas position *within* a geochronologic unit requires a temporal adjective such as: early, middle or medial, late, latest. However, depending on context, temporal adjectives may also be used with chronostratigraphic units, for example, "the youngest part of the system", "the earliest strata of the stage", and so on. (See also Section 7.C.5, p. 73.)

2. Chronozone (and Chron)

a. Definition. *Chronozone* is a formal term for the lowest ranking division in the hierarchy of chronostratigraphic terms (see also Section 7.B.4, p. 67). *Chron* is the corresponding geochronologic term.

b. Time Span. The time span of a chronozone is usually defined in terms of the time span of a previously designated stratigraphic unit such as a formation or a member or a biozone. For example, a formal chronozone based on the time span of a biozone includes all strata equivalent in age to the total maximum time span of that biozone regardless of the presence or absence of fossils diagnostic of the biozone (see Section 7.B.4, p. 67 and Figure 12).

If the unit on which the chronozone is based is of the type that has a designated stratotype (e.g., a lithostratigraphic unit) the time span of the chronozone may be defined in either of two ways: first, it may be made to correspond to the *time span of the stratotype of the unit*; in this case the time span of the chronozone would be permanently fixed. Second, the time span of the chronozone may be made to correspond to the *total time span of the unit* (which may be larger than that of the stratotype); in this case the known time span of the chronozone would vary with increasing information concerning the development of the unit. Where there is an appreciable difference between the time span

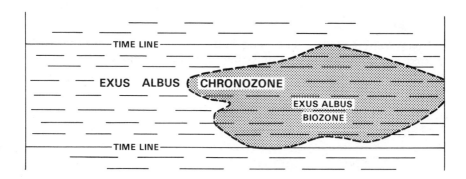

Figure 12. Relation between *Exus albus* Chronozone and *Exus albus* Biozone. (Distribution of specimens of *Exus albus* shown by dot-pattern.)

of the stratotype of the stratigraphic unit and the total known time span of the unit, the definition of the chronozone should be explicit in designating one or the other as the reference, for example, chronozone of the *type* Barrett Formation or chronozone of the Barrett Formation. This is important because while one of the boundaries of a chronozone based on the stratotype of a stratigraphic unit might coincide with one of the boundaries of a stage or substage, the boundaries of a chronozone based on the total time span of a unit will vary in position with changes in information on the time span and diachronism of the unit and hence will not necessarily continue to coincide with the boundaries of the stage or substage although originally made to do so.

If the unit on which a formal chronozone is based is a unit of the type which cannot appropriately have a designated stratotype (such as a biostratigraphic range-zone: taxon-range-zone, concurrent-range-zone, and so on), its time span cannot be defined permanently because the time span of the reference unit may change with increasing information about the time range of the diagnostic taxon or taxons on which the unit is based. Chronostratigraphic units, such as stages, cannot very satisfactorily be subdivided hierarchically into chronozones of this type because, while the time span of a stage is fixed by its boundary-stratotypes, the time span of such a chronozone may vary, not only from place to place, but also in any one place with increasing information about the diagnostic characters of the stratigraphic unit on which it is based. Chronozones of this sort may not only develop overlaps and gaps among themselves, but also their boundaries may not continue to be at the stage boundaries with which they were originally supposed to coincide. For these reasons, stages are better subdivided into substages that can be defined in terms of boundary-stratotypes, and that can be made to cover completely and precisely the time span of the stage to which they belong without gaps or overlaps.

c. Geographic Extent.   The geographic extent of a chronozone is, in theory, worldwide, but its applicability is limited to the area over which its time span can be approximately identified in the strata, which is usually much less.

d. Name.   The chronozone takes its name from the stratigraphic unit on which it is based, for example, *Exus albus* Chronozone (derived from the *Exus albus* Range-zone), Barrett Chronozone (derived from the Barrett Formation). A chron takes the same name as its chronozone.

3. Stage (and Age)

a. Definition.   The *stage* is a chronostratigraphic unit of relatively minor rank in the conventional hierarchy of formal chronostratigraphic terms, representing a relatively minor interval of geologic time. Its geochronologic equivalent is known as an *age* and carries the same proper name as the corresponding stage.

The stage has been called the basic working unit of chronostratigraphy because it is suited in scope and rank to the practical needs and purposes of intraregional chronostratigraphic classification. Furthermore, it is one of the smallest units in the standard chronostratigraphic hierarchy that in prospect may be recognized worldwide.

Stages may be subdivided into substages (Section 7.C.4, p. 72). The relation of stages to chronozones is discussed in Section 7.C.2, p. 70.

b. Boundaries and Stratotypes. Conventionally, the unit-stratotype of a stage should be a single continuously exposed section, in facies favorable for time-correlation, extending from a lower boundary-stratotype to an upper boundary-stratotype. Unfortunately, such single complete sections of strata are not common. Furthermore, the characters of a stage other than its time scope, like those of chronostratigraphic units in general, cannot be defined by any single section no matter how complete (see Section 7.H.2, p. 83). Consequently, a stage is best defined simply by its boundary-stratotypes.

The boundary-stratotypes of a stage should be within sequences of continuous deposition—preferably marine—and both should be associated with distinct marker horizons such as biozone boundaries that can be readily recognized and widely traced as isochronous horizons. The stage boundaries as they are extended away from the boundary-stratotypes should be in principle isochronous. In attempting to determine and extend such isochronous surfaces it is desirable to use as many indicators of time-correlation as possible. For example, it may be desirable to utilize not one but many interfingering biostratigraphic zones.

If major natural changes ("natural breaks") in the historical development of the Earth can be identified at specific points in sequences of continuous deposition, these may constitute desirable points for the boundary-stratotypes of stages. The selection of stage boundaries deserves particular emphasis because they serve to define not only stages but also chronostratigraphic units of higher rank, such as series and systems of which stages are components.

c. Time Span. The lower and upper boundary-stratotypes of a stage represent specific moments in geologic time, and the time interval between them is the time span of the stage. Currently recognized stages are variable in time span, but on the average they range from 3 to 10 million years as indicated by isotopic age determinations. The thicknesses of stages in their type sections may range from a few meters to many thousands of meters. Moreover, the thickness of any one stage may be variable from place to place, depending on the local rate of rock accumulation and the degree of its preservation.

d. Name.   The name of a stage preferably should be derived from a geographic feature in the vicinity of its type section or type area. Conventionally, most stages have been given geographic names. Many currently used stages have the names of the lithostratigraphic units on which they were originally based in their type areas; others have been given names unrelated to other stratigraphic units. In English, the adjectival form of the geographic name is used, with an ending in "ian" or "an," for example, Burdigalian Stage, Cenomanian Stage, Jacksonian Stage, Deiran Stage. An age takes the same name as the corresponding stage.

4.   Substage and Superstage

The *substage* is a subdivision of a stage. Some stages have been divided completely into formally named substages; others have had only certain parts designated as substages. The geochronologic equivalent of a substage might be termed a *subage*, but preferably is called simply an *age*. A *substage* is defined by boundary-stratotypes. Names of substages follow the same rules as those of stages.

Several adjacent stages may be grouped into a *superstage*. However, it is often preferable simply to subdivide the original stage into two or more new stages and, if the need for the original larger unit remains, to make the original stage into a series including the new stages.

5.   Series (and Epoch)

a. Definition.   The *series* is a unit in the conventional chronostratigraphic hierarchy, ranking above a stage and below a system. The geochronologic equivalent of a series is an *epoch*.

A series is always a subdivision of a system; it is usually but not always broken up into stages. Most systems have been divided into three series, but the number varies from two to six. Series commonly include from two to six stages. The terms *superseries* and *subseries* have been used only infrequently, for example, Senonian Subseries (or Superstage) of the Upper Cretaceous Series. Most series can be recognized worldwide, although some as yet have had only more restricted application.

b. Boundaries and Boundary-Stratotypes.   Series are defined by boundary-stratotypes. If a series has been completely divided into stages, its boundaries should be the lower boundary of its oldest stage and the upper boundary of its youngest stage. If stage subdivisions are lacking, the series may be defined independently by its own boundary-stratotypes.

c. Time Span. Currently accepted series vary in time span, but average about 15 million years. The time span of a series, if divided fully into stages, is the sum of the time span of its component stages.

d. Name. A new series name should preferably be derived from a geographic feature in the vicinity of its type section or type area. Nevertheless, the names of currently recognized series which are of other origins should not be changed. Some are geographical, for example, Namurian Series; others are named from their position within a system (lower, middle, upper), for example, Middle Devonian Series, Upper Cretaceous Series; others have Greek word derivations, for example, Miocene. Many of the currently recognized series names of geographic origin have been given the ending "an" or "ian," for example, Chesteran Series, Dinantian Series.

The epoch corresponding to a series takes the same name as the series except that the terms lower, middle, and upper applied to a series are changed to early, middle, and late when referring to an epoch. In both cases, these terms are capitalized when referring to a formal unit, for example, Lower Devonian, Early Devonian; but not for informal reference to chronostratigraphic or geochronologic position, for example, "in the lower part of the Devonian" and "early in Devonian time."

e. Misuse of "Series". The term series has been frequently used incorrectly as a lithostratigraphic term more or less equivalent to a group and consisting of an alternation of lithologic types. This usage should be discontinued.

6. System (and Period)

a. Definition. The *system* is a unit of major rank in the conventional chronostratigraphic hierarchy, above a series and below an erathem. All of the generally accepted systems have a time span sufficiently great so that they serve as worldwide chronostratigraphic reference units. In fact, of all units of the chronostratigraphic scale, they are probably the most widely recognized and the most widely used to indicate general chronostratigraphic position. The geochronologic equivalent of a system is a *period*.

Special circumstances have suggested the occasional need for *subsys-*

tems and *supersystems*, for example, Mississippian Subsystem of the
Carboniferous System, Karroo Supersystem.

b. Boundaries and Boundary-Stratotypes.   As in the case of stages and
series (Sections 7.C.3, 71, and 7.C.5 p. 72-73), the boundaries of a system
are defined by boundary-stratotypes. If a system is divided into series and
stages, its lower boundary-stratotype is that of its oldest series or stage and
its upper boundary-stratotype is that of its youngest series or stage.

The boundaries of most recognized systems are at present ill defined,
uncertain, and controversial in varying degrees. This is due to inexact
original definitions, to the discovery of gaps and overlaps at the level of
what were previously thought to be the boundaries of adjacent systems,
and to the lack of a universally accepted concept of what a system is
and how its boundaries should be defined.

A primary step in refining the definition of a system is to decide just
what stages and series are to be included in the system. The definition
of these component stages and series then automatically defines the
system and its boundaries.

The procedure for extending system boundaries geographically away
from the type is the same as that for extending other chronostratigraphic horizons (see Section 7.J, p. 86-92).

c. Time Span.   The time span of a system can most readily be defined
as the time span of the sum of its component series or component
stages. The time spans of the currently accepted Phanerozoic systems
range from 35 to 70 million years and average about 50 to 60 million
years.

d. Name.   The names of currently recognized systems are of diverse
origin. Some are indicative of position (e.g., Tertiary and Quaternary);
others have a lithologic connotation (e.g., Carboniferous, Cretaceous);
others are tribal (e.g., Ordovician, Silurian); and still others are geographic (e.g., Permian, Devonian). Likewise, they also bear a variety of
endings, such as "an," "ic," and "ous." There is no pressing need to
standardize the derivation of system names. The period takes the same
name as the system to which it corresponds.

Certain stratigraphic units in parts of the world distant from western
Europe are locally called "systems," although they do not coincide with
the so-called standard systems and are somewhat larger in scope. Such
are the Karroo "System" of Africa and the Hokonui "System" of New
Zealand. If these are used in a chronostratigraphic sense, they might be
considered informal systems or supersystems.

7. Erathem (and Era)

An *erathem* (from the Greek roots *era* and *them*, "the deposit of an era") is the largest formal unit commonly recognized in the chronostratigraphic hierarchy and usually consists of several adjacent systems. The interval of geologic time corresponding to an erathem is an *era*. It carries the same name as its equivalent erathem.

Erathems have traditionally been named to reflect major changes in the development of life on the Earth: Paleozoic (old life), Mesozoic (intermediate life), and Cenozoic (recent life). The term Precambrian is now most commonly used for rocks older than Paleozoic, although Archeozoic (most ancient life) is also used by some and has the advantage of clearly indicating a relationship to the development of life and thus paralleling in derivation the other "zoic" era terms. However, considering the immense time span represented by pre-Paleozoic rocks, it is questionable whether these rocks should be assigned to any single erathem. It seems preferable to think of them as belonging to an eonothem (see Section 7.C.8), which may eventually be broken down into several erathems.

8. Eonothem (and Eon)

The term *eon* has been used for a geochronologic unit greater than an era. Logically, the chronostratigraphic equivalent would be an *eonothem*. Two eons are generally recognized. One is the Phanerozoic eon (time of evident life), which encompasses the Paleozoic, Mesozoic, and Cenozoic eras. The other eon covers pre-Phanerozoic (pre-Paleozoic) time and has been known as the Cryptozoic Eon (time of hidden life), the Archeozoic Eon (time of most ancient life), or simply the Precambrian Eon. The last term is most commonly and widely used, although it has the disadvantage of suggesting for its younger counterpart the awkward term of "post-Precambrian." The situation has been further complicated by the widespread reference to an "Infracambrian" or "Eocambrian" unit, after the true Precambrian but before the Cambrian. Both terminology and nomenclature of time and rocks before the Cambrian remain unsettled (see Section 7.D.2, p. 77, 81-82).

9. Informal Chronostratigraphic Units

Many formal chronostratigraphic terms and their geochronologic equivalent are also used informally, for example, chronozone of the dinosaurs, the age of mammals, a period of deposition. In formal usage in English, named terms should always be written with an initial capital letter, whereas in informal usage, terms should follow the same rules of capitalization as do ordinary common nouns.

The concept of the informal chronozone is particularly valuable to express the total body of strata equivalent in age to *any* feature having a stratigraphic range in time, for example, the chronozone of the ammonites, the *Globotruncana* chronozone, the Brunswick chronozone (all rocks of the same age as the Brunswick Formation), the chronozone of the Olduvai magnetic reversal. It is also useful as a temporary approach to the eventual establishment of formal chronostratigraphic units in unexplored regions such as the oceans. In current studies of the stratigraphy of suboceanic strata from drill holes, for instance, attempts to determine the informal chronozones of various lithostratigraphic, biostratigraphic, paleomagnetic, and other kinds of units have provided the necessary background for the eventual attainment of a few reliable formal chronozones or stage subdivisions.

D. STANDARD GLOBAL CHRONOSTRATIGRAPHIC (GEOCHRONOLOGIC) SCALE

1. Concept

As mentioned above, a major goal of chronostratigraphic classification is the establishment of a hierarchy of chronostratigraphic units of worldwide scope, which could serve as a standard scale of reference for the dating of all rocks everywhere and for relating all rocks everywhere to world geologic history (see Section 7.A, p. 66).

All units of the standard chronostratigraphic hierarchy are theoretically worldwide in extent, as is their corresponding time span. However, only the units of higher rank in this hierarchy lend themselves at present to worldwide application. The effective geographic extent of chronostratigraphic units decreases with decreasing rank of the unit because of limitations in the resolving power of long-range time-correlation away from the stratotypes. Thus systems are generally recognized worldwide; series usually so; but units of lower rank are at present commonly of only regional or local application, although their recognition worldwide is a goal.

2. Present Status

Table 3 shows a Global Chronostratigraphic (Geochronologic) Scale of common current usage (though not universally accepted) to which are added approximate isotopic ages and durations of periods in millions of years. Such a scale should be extended to include standard series (epochs) and stages (ages). There are, however, many questions and problems concerning the elaboration of such a scale. The notes on Table 3 suggest only a few of these questions and problems.

There is, for instance, considerable controversy as to the *named* units that should be recognized for such a scale, even for the major worldwide ranks such as eras and erathems, and periods and systems. The boundaries of nearly every system are in dispute; the scope of systems and series in terms of stages is controversial; many of the systems have been subdivided into units which some consider to be series and others consider to be stages; and so on.

In the Precambrian, rocks representing intervals many times greater than those of Phanerozoic periods still usually can be assigned to units of only local or regional extent.

3. Recommendations for Definition of Units

In recent years there has been growing interest in clarifying the scope of each of the Phanerozoic systems and in establishing stratotype standards for the boundaries between them. The IUGS International Commission on Stratigraphy has established subcommissions for most of these systems, and the subcommissions on adjacent systems have formed or are forming joint working groups to study and make recommendations for fixing the intervening system boundaries.

Principles, problems, and procedures with respect to system boundaries were discussed in ISSC Report-2 (1964), *Definition of Geologic Systems*, which strongly recommended the establishment of type boundary points in continuously deposited sections as the best means of providing uniform standard definitions for the systems and their principal hierarchical subdivisions. These recommendations were further elaborated in ISSC Report-4, *Stratotypes* (1970) and ISSC Report-6, *Chronostratigraphic Units* (1971).

A beginning in the application of these recommended procedures for the establishment of boundary-stratotypes for the global systems was made in 1972 when agreement was reached by the IUGS Commission on Stratigraphy's Working Group on the Silurian-Devonian Boundary to establish a global standard boundary-stratotype for the Silurian-Devonian boundary in the section at Klonk, Czechoslovakia, at a point coinciding with the apparent base of the *Monograptus uniformis* Range-zone *in that section*. It is to be hoped that global standard boundary-stratotypes between other systems and their principal subdivisions will be established in a similar manner in the near future.

Table 3. Major Units of Standard Global Chronostratigraphic
(Geochronologic) Scale

EONOTHEMS AND EONS[a]	ERATHEMS AND ERAS	SYSTEMS AND PERIODS[b]	ISOTOPIC DATING in millions of years[c]	
			DURATION OF UNIT	AGE OF BEGINNING OF UNIT
Phanerozoic	Cenozoic[d]	Quaternary[e]	2	2
		Tertiary[f]	62	64
	Mesozoic	Cretaceous	76	140
		Jurassic	68	208
		Triassic	34	242
	Paleozoic	Permian	42	284
		Carboniferous[g]	76	360
		Devonian	49	409
		Silurian	27	436
		Ordovician	64	500
		Cambrian	64	564
Archeozoic[h] Cryptozoic[i] Precambrian[j]			3000+	3700+

[a] The Paleozoic, Mesozoic, and Cenozoic Eras are sometimes grouped together to
form the Phanerozoic Eon, as contrasted with another older eon, representing 85
percent of geologic time, known variously as Precambrian, Cryptozoic, or Archeozoic Eon. The term Precambrian is the most widely used.
[b] The systems recognized by the International Geological Congress in Paris (1900)
were: Modern, Tertiary, Cretacic, Jurassic, Triassic, Carbonic, Devonic, Siluric,
and Cambric. However, Modern has not succeeded in replacing Quaternary; a
Permian System is now generally recognized; the name Ordovician System was
officially accepted by the Norden Congress (1960) for the older part of the original
Silurian System; and the *ic* ending is used at present only for Triassic and Jurassic.
[c] Approximate values based on a review of existing data presented by R. L.

Armstrong at the meeting of the IUGS Subcommission on Geochronology in Paris, August 1974.

[d] Sometimes written Kainozoic or in inconsistent transcription, Cainozoic.

[e] The names Quaternary and Tertiary are anachronisms because the antecedent Primary and Secondary are no longer used as system names. The name Anthropogene is used by some to replace Quaternary. The Quaternary System is divided into two series: a younger Holocene Series and an older Pleistocene Series. The duration of the series of the Quaternary is incongruously short as compared with that of series of other systems.

[f] Instead of a Tertiary System, many recognize a Neogene System (Miocene and Pliocene Series) and a Paleogene System (Paleocene, Eocene, and Oligocene Series).

[g] In North America, in place of a Carboniferous System, two systems are commonly recognized: a Mississippian System (older) and a Pennsylvanian System (younger). These are also sometimes known as subsystems of the Carboniferous System.

[h] The name Archeozoic appears to have been proposed first by James Dwight Dana in 1872 for the initial era of geological history, preceding Paleozoic time. The word is derived from the Greek *archeo* (most ancient, primitive, beginning) and *zoic* (life). It is thus very appropriate in meaning and derivation to include all strata and time older than Paleozoic; in view of remains of early life found in strata 3100 million years old and the character of the earliest sediments, the origin of life may be essentially as old as the Earth's oldest known rocks (±3700 million years). The name has been used with both "era" and "eon." It has fallen into disuse because of unsuccessful attempts to use it for only the oldest division of the Precambrian, but if restored to its original definition and raised to the rank of an eon, it would be quite satisfactory. A similar and even older candidate is the name Protozoic (Murchison, 1838), but this name is less desirable because it might be confused with Proterozoic.

[i] The name Cryptozoic (hidden life) was proposed in 1930 by G. H. Chadwick (in a published abstract of a never-completed paper), as a counterpart to Phanerozoic and an escape from what he considered the "hopelessly negative" expression "pre-Cambrian." He translates *crypto* as "obscure" and *zoic* as "animal" and questions whether Cryptobiotic might not be preferable. The expression "Cryptozoic Eon" subsequently has gained considerable usage for pre-Phanerozoic (pre-Paleozoic) time.

[j] The name Precambrian seems to have developed from the repeated use of "pre-Cambrian." It is a rather clumsy and not very appropriate name but is currently by far the most widely used name for time and rocks older than Paleozoic. It has been used both as an era (erathem) and an eon (eonothem), but the latter seems preferable. Many attempts have been made to subdivide it into eras (erathems), but few of such proposed divisions have been adequately established or widely accepted. An upper division, Proterozoic, and a lower division, Archean, have been widely used but a consistent relative scope for the two has never been satisfactorily established.

(It should be noted that the Silurian-Devonian boundary worldwide was *not* defined as the base of the *M. uniformis* Range-zone, but was simply fixed at a time-horizon coinciding with a point in the section at Klonk marked as the boundary-stratotype. To the extent of present knowledge, the *M. uniformis* Range-zone in the Klonk section has its base at this boundary-stratotype point, and the base of the *M. uniformis* Range-zone thus becomes a useful general *guide* to this boundary. However, the base of the zone elsewhere will not necessarily coincide exactly with this horizon at Klonk, and it is always possible that even in the Klonk section future discoveries may extend the range of *M. uniformis* lower than the boundary-stratotype.)

Principal points in the ISSC recommended procedure for the definition of systems (or other units) of the Standard Global Chronostratigraphic Scale as summarized from its previous Reports (see also Section 7.H, p. 82-86), are:

1. Organization of a working group of the IUGS Commission on Stratigraphy composed of a wide geographic spread of members particularly knowledgeable about the part of the stratigraphic column under consideration, supplemented by correspondents from pertinent specialized fields or local regions.

2. Study and review of the concepts and historical background of the adjacent systems and previous and current attempts at boundary definition between them.

3. Worldwide review of the geographic distribution of the adjacent systems and identification of areas and sections where strata in general proximity to their mutual boundary are well exposed and accessible.

4. Review of potential widespread correlation horizons in the general boundary zone between the two systems and their probable value for regional or global time-correlation—biozones and biohorizons, lithologic features, magnetic reversals and other paleomagnetic features, isotopic age determinations, eustatic changes of sea level, major unconformities, orogenies, paleoclimatic changes, and such.

5. Selection of specific sections for study and consideration based on probable continuity of sedimentation through the critical boundary interval; completeness of exposure; adequate thickness of sediments; abundance and variety of well-preserved fossils; favorable facies for development of widespread, reliable, and time-significant correlation horizons; close ties to other facies; freedom from structural complication, metamorphism, or other alteration; freedom from unconformities; amenability to isotopic age determination; historical appropriateness; and accessibility.

6. Field, laboratory, and literature studies of the most favored sections based on the criteria outlined above.

7. Decision by the working group as to the best stratotype section.

8. Selection in the field of the precise position of the boundary-stratotype in the chosen section, so as to best express the appropriate concepts of the two adjacent systems, and so as to be most practicably correlatable as an approximately isochronous horizon worldwide.

9. Approval by the IUGS Commission on Stratigraphy, and the IUGS, of the boundary stratotype as the global standard for the boundary between the two systems.

10. Marking of the boundary-stratotype in the field and establishment of arrangements for its preservation and accessibility to authorized study.

E. REGIONAL CHRONOSTRATIGRAPHIC SCALES

The units of a Standard Global Chronostratigraphic Scale are valid only as they are based on sound, detailed local and regional stratigraphy. Accordingly, the route toward recognition of uniform global units is by means of local or regional stratigraphic scales, particularly with respect to chronozones, stages, and series. Moreover, regional units of this rank will probably always be needed whether they fit exactly into the standard global units. It is better to refer strata with accuracy to local or regional units rather than to strain beyond the current limits of time-correlation in assigning these strata to units of a global scale.

F. CLASSIFICATION OF THE PRECAMBRIAN

The rock record of the Precambrian, representing perhaps as much as 85 percent of geologic time, has not yet been divided systematically into globally recognized chronostratigraphic units. However, there are prospects that chronostratigraphic classification of much of the Precambrian may eventually be attained through isotopic dating and through such other means as lithologic sequence, stromatolites, paleomagnetic signature, and relation to volcanic or plutonic episodes, orogenic cycles, major climatic changes, geochemical events, and major unconformities. The basic principles to be used in dividing the Precambrian into major chronostratigraphic units should be the same as for Phanerozoic rocks, even though different emphasis may be placed on the various means used for time-correlation.

As in the Phanerozoic, the definition of chronostratigraphic units in the Precambrian as intervals between designated points in the rock sequence (boundary-stratotypes) leaves the way open to use all methods of time-correlation; and although in the Precambrian heavy reliance will be placed on isotopic dating, there should remain the fixed standard of definition *in the rocks* for these units—one that will allow the use of all methods known at present, as well as new ones to come, for the extension and identification of the units.

In the Precambrian, as in the Phanerozoic, the logical procedure is first to build up local chronostratigraphy in appropriate areas, utilizing all possible guides to local time-correlation; then to proceed geographically from local to regional to continental to worldwide scope, as the means and evidence justify. Local chronozones of whatever rank, defined by boundary-stratotypes, will furnish units useful for local Precambrian history regardless of any worldwide scheme. They will also constitute the best possible foundation for regional, continental, and worldwide units if, as, and when these can be established with reasonable assurance.

This classification of the Precambrian may be usefully supplemented by a scheme of chronometric units based on isotopic dating. However, such units may vary with corrections or change in isotopic determinations and cannot be considered to have as stable a base as chronostratigraphic units defined by boundary-stratotypes.

G. QUATERNARY CHRONOSTRATIGRAPHIC UNITS

The basic principles to be used in dividing the Quaternary into chronostratigraphic units should be the same as for the other Phanerozoic strata, although different emphasis may be placed on the various means (climatic, magnetic, isotopic, etc.) used for time-correlation. Carbon-14 dating has been particularly useful in the late Quaternary.

Although it may often be impracticable to establish continuous type sections or comprehensive local unit-stratotypes for Pleistocene and Holocene chronostratigraphic units, the characterization of such units as the intervals *between* certain designated boundary-stratotypes would seem to be the best means for their definition.

H. PROCEDURES FOR ESTABLISHING CHRONOSTRATIGRAPHIC UNITS

See also Section 3.B, p. 16-17 and Section 7.D.3, p. 80-81.

1. Stratotypes as Standards

Each named chronostratigraphic unit of whatever rank should have a clear, constant, and precise standard definition that will always mean the same thing to everyone everywhere. The essential part of such a definition is the time span of the unit described. Since the only record of geologic time and of the events of geologic history lies in the rock strata themselves, the best standard for the definition of a chronostratigraphic unit is a specifically designated stratigraphic interval—unit stratotype—between two designated reference points—lower and upper *boundary-stratotypes.*

2. Unit-Stratotypes

The unit-stratotype of a chronostratigraphic unit is conventionally considered to be a reasonably continuous, designated section through the entire unit in its type area, a section that exposes particularly well its lower and upper limits (boundary-stratotypes) and thus defines the essential feature of the unit—its time span. The only true unit-stratotype for the internal physical characters (lithology, fossil content, etc.) of a chronostratigraphic unit would be the sum total of all possible sections through the unit, representing all facies in which it is developed. These internal physical characters of a chronostratigraphic unit, while not diagnostic of the unit, are very important in facilitating correlation of the unit and its extension and recognition in other areas. However, they play no part in defining the essential time scope of the unit which depends solely on the position of the boundary-stratotypes.

In sections used as unit-stratotypes, it is desirable to have as few gaps (covered intervals, diastems, unconformities, structural cutouts, etc.) as possible. Although these internal breaks do not affect the definition of the unit's time span, they may make its extension to other areas more difficult. Designation of reference sections (hypostratotypes) in many different geographic areas amplifies the concept of the unit and helps in extending the concept of the unit away from the type area.

3. Boundary-Stratotypes

Stratotypes of the lower and upper boundaries of a chronostratigraphic unit best define its time span, which is its diagnostic character. The two boundary-stratotypes need not be a part of a single unit-stratotype section, nor need they be in the same locality. Both, however, should be chosen in sequences of essentially continuous deposition, even if this means establishing them *within* individual beds, since the reference points for the boundaries should represent points in time as specific as possible. The worst possible boundary is an unconformity; it not only does not represent a sharp point in time but also tends to change in age laterally.

In areas where strata are exposed overlying a major regional angular unconformity (e.g., Mesozoic sediments transgressing on a Precambrian erosion surface), the usual practice has been to identify the lower boundary-stratotype of the overlying unit as a point at the intersection between the unconformity, and the base of the oldest known beds overlying the unconformity. The unconformity surface thus became a convenient physical limit of the unit in the area. This procedure is valid only if the point identified as marking the base of the unit at the boundary-stratotype is clearly recognized as the stratigraphic horizon that abuts the unconformity and *not* the unconformity itself. If additional strata are found elsewhere below this horizon, but still above the unconformity, they must then be accommodated in a different and older chronostratigraphic unit.

The boundary-stratotypes of chronostratigraphic units should be selected at or near markers favorable for long distance time-correlation. Commonly, they are made to coincide with a boundary of some biostratigraphic or lithostratigraphic unit. Particularly useful points for boundary-stratotypes of chronostratigraphic units include biostratigraphic horizons in marine sequences with abundant planktic fossils, points that can be dated accurately by radiometric determinations, and points of magnetic reversal.

4. Advantage of Defining Chronostratigraphic Units by Mutual Boundary-Stratotypes

Chronostratigraphic units should ideally be defined by adequately established stratotypes, with the units of each rank in the hierarchy completely filling, without overlap, the appropriate unit of the next higher rank. Each rank would thus be made up of a single set of subdivisional units including in total all strata found in the whole time interval corresponding to the unit of the next higher rank. This could readily be attained if the whole sequence of strata representing all geologic time were completely exposed in some one section, or if means of time-correlation were always so good that the horizon marking the top of one unit in its type locality could be identified with assurance as the base of the next succeeding unit whose type locality might be in some other area. Neither of these conditions exists and this poses a problem.

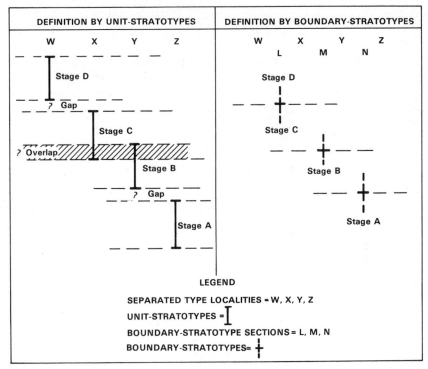

Figure 13. Advantage of defining stages by mutual boundary-stratotypes, where type localities are widely separated, rather than by unit-stratotypes.

As an example, a stage may have its type locality in one area and the adjacent under- or overlying stages may have their type localities in other areas (left-hand side of Figure 13). In such cases, there is a problem in making certain that the top boundary of the unit-stratotype of one stage corresponds exactly with the bottom boundary of the unit-stratotype of the next younger stage. Correlation of the boundary between two successive stages from the type area of one stage to that of the other is usually not so good but that there is danger of gaps or overlaps between the type limits of the two successive stages. For these reasons it is preferable to select a single common (mutual) boundary-stratotype to serve both as the top of one stage and the bottom of the next younger stage (right-hand side of Figure 13). This practice guarantees that the two type boundaries are identical and obviates difficult boundary correlations between distant areas. At the same time it allows the type exposures of both sequentially adjacent units to be those of their respective type areas. It also allows the use of complete unit-stratotypes for those units whose boundary-stratotypes follow each other in sequence in the same type area.

The boundary-stratotypes between stages could be selected so that certain ones could serve also as the boundary-stratotypes between larger units (series, systems, etc.). The procedure thus lends itself readily to a complete hierarchical scheme of chronostratigraphic divisions with no gaps and no overlaps.

J. MEANS FOR EXTENDING CHRONOSTRATIGRAPHIC UNITS—CHRONOCORRELATION (TIME-CORRELATION)

Only after the type limits (boundary-stratotypes) of a chrono-stratigraphic unit have been established can the limits be extended geographically beyond the type section. The boundaries of a chronostratigraphic unit are by definition isochronous surfaces, so that the unit everywhere includes rocks of the same age. In practice, the boundaries are isochronous only so far as the resolving power of existing methods of time-correlation can prove them to be so. Generally, an ideal of isochroneity can be approached only with diminishing accuracy as the chronostratigraphic boundaries are followed at increasing distance from the type. Therefore, all possible lines of evidence should be utilized for time-correlation (chronocorrelation): distribution of fossils of many kinds, trace of beds, sequence of beds, lithology, isotopic dating, electric log markers, unconformities, transgressions and regressions, volcanic activity, tectonic episodes, paleoclimatology, paleomagnetic signature, and so on. Nevertheless, the isochronous boundaries of chronostratigraphic units are inherently independent of all other kinds of stratigraphic boundaries, except as these may serve as local guides to chronostratigraphic position.

1. Physical Interrelations of Strata

The simplest and most obvious clue to the relative age or chrono-stratigraphic position of rock strata lies in their physical interrelations. The classic Law of Superposition of Strata simply states that in an undisturbed sequence of sedimentary strata the uppermost strata are younger than those on which they rest.

The order of superposition of strata provides the most unequivocal indicator available for relative age relations, and it is healthy to recall that all other methods of age determination, both relative and absolute, were originally dependent, directly or indirectly, on observed physical sequence of strata as a check and control on their validity. For a sufficiently limited distance, the trace of a bedding plane is frequently the best indicator of isochroneity.

Difficulties arise, however, when strata are severely disturbed, overturned, or overthrust; when a younger igneous rock is emplaced within a sequence of older strata; when a relatively mobile sedimentary rock like shale, salt, or gypsum is diapirically injected into younger strata or flows over them; or, perhaps most importantly, when continuous exposures are lacking and there are discontinuities due to lateral variability, overlap, unconformities, faulting, intrusion, and so on. Even in such difficult situations, correlation by means of physical character and stratigraphic sequence is almost always helpful in relative age determination.

2. Lithology

At one time many of the systems and their subdivisions were primarily lithostratigraphic divisions whose distinctive lithology was supposed everywhere to characterize the rocks generated during certain intervals of geologic time. However, it was soon recognized that lithologic character is commonly influenced more strongly by environment than by age, that the boundaries of all lithostratigraphic units eventually cut across isochronous surfaces and vice versa, and that lithologic features are repeated time and again in the stratigraphic sequence. Even so, a lithostratigraphic unit such as a formation always has some chronostratigraphic connotation and is useful as an approximate guide to chronostratigraphic position. Individual limestone beds, phosphate beds, bentonites, volcanic ash beds, or tonsteins, for example, may be excellent guides to approximate time-correlation over very extensive areas. Distinctive and widespread general lithologic developments also may be significant of chronostratigraphic position.

3. Paleontology

Because of their highly distinctive character, fossils constitute one of the best and most widely used means of tracing and correlating beds and thus determining the relative age of one to another. Moreover, the rather orderly and progressive change of fossils with time as a result of biologic evolution furnishes an independent and impressively effective key to the age and relative position of strata worldwide. Because the course of organic evolution is irreversible with respect to geologic time and the remains of life are widespread and distinctive, fossils have indeed constituted the best means for worldwide relative dating and approximate long-distance time-correlation, throughout the younger part of the Earth's geologic column, and have largely made possible the development to date of a global chronostratigraphic time scale for Phanerozoic strata.

Although biostratigraphic correlation is not necessarily time-correlation, it has been and continues to be one of the most useful approaches to time-correlation if used with discretion and judgment. Biostratigraphic methods are continually being refined to make them increasingly effective. Two fossiliferous intervals at widely separated localities may have large differences in general fossil content because of facies changes, but subtle paleontological discrimination may show that there is a time-correlation between them. On the other hand, two superficially similar fossil assemblages may similarly be shown to be of quite different ages.

Although no individual biozone has either a lower or an upper boundary which is everywhere of identical age, the use of *several interlocking* biozones, laterally interfingering and replacing each other, may often provide a reasonably assured indication of approximate isochronous position. Such a system of interlocking biozones can be particularly helpful in providing a tie across major lateral changes in depositional environment. An example is the use of the land-to-ocean progression of terrestrial animals and plants, pollen, benthic marine organisms, and planktic and nektic marine organisms in the correlation between continental and marine deposits. Another example is the use of overlapping plant and animal zones in correlation from tropical to temperate to polar environments.

Another effective paleontological key to long-range time-correlation is through the restoration of evolutionary sequences of fossil forms and the use of interlocking lineage zones. Numerous statistical techniques have been developed for this purpose.

A realization of the problems to be faced in paleontological time-correlation may be gained from consideration of the variety of life-environments on Earth at the present time and the great lateral variation in living forms. With the added complexities due to fluctuating environments of the past, continental drift, diagenetic changes in strata, metamorphism, vagaries of fossil preservation, time required for migration, accidents of collection, and other factors (see Figure 14), it is understandable that along with its great value, long-range paleontological time-correlation also has serious limitations. Moreover, Precambrian rocks, constituting a large part of the Earth's crust and corresponding to 85 percent of geologic time, are largely lacking in usable fossils, and even in the Phanerozoic not all strata are fossiliferous, and fossils where present yield only relative age, not age in years.

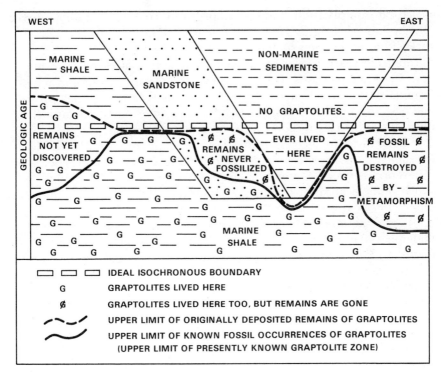

Figure 14. Possible causes of local variation in the relation of both the original upper limit of occurrence of a graptolite taxon and the upper limit of present known occurrences of the taxon to an isochronous horizon (chronostratigraphic horizon).

4. Isotopic Age Determinations

Another outstanding key to chronostratigraphy is provided by isotopic dating methods based on the radiometric decay of certain parent isotopes at a rate that is not only constant but also suitable for measuring geologic time. The most commonly used methods (U-Pb, Rb-Sr, K-Ar) produce data with high precision—analytical errors in the range of 0.1 to 2 percent.

Isotopic dating not only allows relative dating of strata but also is almost unique in its potentiality for contributing age values expressed in years or millions of years. It has furnished the first reasonably reliable quantitative evidence of the length of geologic time, suggesting that the age of the oldest known rocks of the Earth's crust is at least 3700 million years. Isotopic dating has also provided the major hope of working out to some extent the ages and age relationships of the great mass of Precambrian strata, where fossils are less effective and where structural complication and metamorphism frequently inhibit direct observation of original stratal sequences. Likewise for Phanerozoic rocks, isotopic age determinations now provide useful data on ages and durations in years, as well as valuable checks on relative age determinations by other methods. Under certain circumstances, isotopic age determinations of intrusive or extrusive igneous rock bodies may provide the best, or even the only, basis for age determination and chronostratigraphic classification of sedimentary sequences.

Discrepancies in age results may arise from the use of different decay constants. This is especially the case for the Rb-Sr system where the difference between commonly used values is about 5 percent. For geochronological comparisons it is therefore important that uniform sets of decay constants be used in the presentation of ages.

Isotopic methods may be applied both to whole rock samples and to minerals separated from these rocks. The errors in physical measurements are small and well under control, but the age significance of isotope data depends on a variety of geological parameters and the use of isotopic methods in chronostratigraphy generally requires geological interpretation. The various isotope systems in different mineral and rock samples may reflect a specific response to varying conditions of pressure, temperature, or other vicissitudes which they have experienced. Thus it may be necessary to decide whether the age obtained is that of metamorphism or other subsequent alteration—rather than the true age of origin of the strata. Likewise, detrital minerals derived from such older sources may cause erroneous conclusions on the age of origin of a rock stratum. Finally, an important limitation to use is that not all rock types are amenable to isotopic age analysis.

A method of age determination through radioactivity differing from those mentioned above is that based on the proportion of the radiocarbon isotope (C^{14}) to normal carbon in the organic matter of sediments. This method has been extremely valuable but is limited in application to the dating of late Quaternary strata.

5. Geomagnetic Reversals

The phenomenon of periodic reversals of the Earth's magnetic field is utilized importantly in chronostratigraphy, particularly in Cenozoic and late Mesozoic strata, where a magnetic time scale is being developed. It has been especially useful in the late Tertiary and Quaternary, where a more detailed chronostratigraphic classification has been developed than is permitted by the resolving power of biologic evolution. Also it is playing an important role in determining the chronostratigraphy of oceanic regions.

6. Paleoclimatic Changes

Climatic changes leave a conspicuous imprint on the geologic record in the forms of glacial deposits, evaporites, red beds, coal deposits, paleontological changes, and such. Since many climatic changes appear to have been regional or worldwide, their effects on the rocks provide valuable information for chronocorrelation. The extent of their effects is complicated, however, by normal variations in climate due to latitude, elevation, oceanic circulation, plate movements, and other factors.

7. Paleogeography and Eustatic Changes in Sea Level

Alternating transgressions and regressions of the sea and the resulting unconformities have classically provided the basis for local and regional division of stratal sequences and many of the chronostratigraphic units of Western Europe originated in this way. As a result of either epeirogenic movements of the land masses or eustatic rises and lowerings of the sea, certain periods of Earth history seem to have been characterized worldwide by a general high or low stand of the continents with respect to sea level. If the level of the sea did indeed rise and fall periodically during geologic time, the evidence in the rock sequence of these eustatic changes can furnish an excellent basis for establishing a worldwide "natural" chronostratigraphic framework. However, local vertical movements of the Earth's crust may have been so great and so variable geographically that the record in the rocks may be difficult to interpret locally.

8. Orogenies

A classic concept of historical geology is that periodic worldwide orogenies have furnished "natural" worldwide dividing lines in Earth history and that these can be identified in the rock strata by their effects on sedimentation, erosion, igneous activity, and rock deformation. This is indeed confirmed for certain regions; and, to some extent, general times of crustal disturbance stand out worldwide, as reflected in the usage of such terms as Caledonian, Hercynian, Nevadan, Laramide, and Alpine orogenies. In the record of isotopic dating there is considerable support for broad cyclic periods of crustal metamorphism. Particularly in the Precambrian, chronostratigraphic classification has been attempted on the basis of worldwide orogenic cycles. However, the considerable duration of many orogenies, their local rather than worldwide nature, their lack of coincidence with classic system or series boundaries, and the difficulty of identifying them closely make them generally unsatisfactory indicators of worldwide chronostratigraphic units.

9. Unconformities

Many of the geologic systems were originally defined as representing
the rocks lying between certain major unconformities because these
appeared to mark natural breaks in lithology, paleontology, and other
features. However, a surface of unconformity inevitably varies in age
and in time-value from place to place and is never universal in extent.
Moreover, unconformity frequently results from very slow epeirogenic
movements taking place over long periods of geologic time. Hence
while unconformities frequently serve as useful guides to the approximate placement of chronostratigraphic boundaries, they cannot in
themselves fulfill the requirements of such boundaries (see also Section
7.H.3, 2nd paragraph, p. 84).

Although unconformity surfaces are not isochronous and continually
cut across time horizons, major regional unconformities obviously have
very   important,   though   broad,   time   significance.   Likewise,
unconformity-bounded units—*synthems*—form a class of stratigraphic
units which, though not chronostratigraphic, have great significance in
chronostratigraphy.

10. Other Indicators

Many other lines of evidence may under limited circumstances be
helpful as guides to time-correlation and as indicators of chronostratigraphic position. For example, certain invertebrates may provide a
valuable clue to chronostratigraphic position because they show a decreasing number of daily growth increments per year due to slowing of
the Earth's rate of rotation in response to tidal impedance.

Various mineralogical, geochemical, and geophysical features of rock
strata provide means of approximate time-correlation over considerable distances. Detrital heavy-mineral assemblages are valuable in
time-correlation and in determining relative time of origin. Varves and
seasonal bands in sediments are indicators of age and duration of
stratigraphic intervals. Probable rates of sedimentation are indicators of
the time required for the formation of sedimentary sequences. Seismic
profiles and electrical and nuclear logging of boreholes provide very
useful means of time-correlation and detailed evidence of relative
chronostratigraphic position. Several special isotopic methods not mentioned above have been developed for dating very young sediments.
Various   dating   methods   have   been   tried   utilizing   thermoluminescence, radiation tracks, pleochroic halos, and other forms of
radiation damage. Many other means could be mentioned, and it is
expected that many totally new methods will be developed.

Many of the above-mentioned contributors to time-correlation, though of only limited accuracy, can be useful in working out the time-relations of strata under the right circumstances. Some are more used than others but none should be rejected. Even with the help of all, time-correlation to extend the boundaries of chronostratigraphic units geographically away from their type areas is never as accurate as could be desired.

K. NAMING OF CHRONOSTRATIGRAPHIC UNITS

A formal chronostratigraphic unit should be given a binomial designation—a proper name plus a term-word—and the initial letters of both should be capitalized, for example, Cretaceous System. The geochronologic equivalent of a formal chronostratigraphic unit should use the same proper name combined with the equivalent geo-chronologic term, for example, Cretaceous Period. The proper name of a chronostratigraphic unit may be used alone where there is no danger of confusion, for example, "the Aquitanian" in place of "the Aquitanian Stage."

Conventions for the names of individual kinds or ranks of chrono-stratigraphic units are discussed under the headings of the units. Chronostratigraphic nomenclature follows the general rules for strati-graphic nomenclature given in Chapter 3.

L. REVISION OF CHRONOSTRATIGRAPHIC UNITS

Much confusion concerning the scope of specific chronostratigraphic units has arisen due to inadequate definition of the units at the time they were proposed. To make these units more useful, it is urged strongly that originally inadequate definitions of units now in common use should be revised to accord with recommended procedures. It is also urged that any new chronostratigraphic unit for which formal status is desired should be adequately proposed and defined according to the procedure outlined in Section 3.B, p. 16-17.

Chapter Eight
Relation Between Litho-, Bio-, Chrono-, and Other Kinds of Stratigraphic Units

The various categories of stratigraphic classification are all closely related in that all deal with rocks of the Earth *as strata*, with the picture of the stratified Earth as it presently exists, and with the history of the Earth as interpreted from its rock strata. Each category, however, is concerned with a different property or attribute of strata and a different aspect of Earth history. The relative importance of the different categories varies with circumstances. Each is important for particular purposes.

Lithostratigraphic and *biostratigraphic* classifications are usually early steps in working out the stratigraphy of a region. When dealing with unfossiliferous rocks, lithostratigraphy is the main initial approach to stratigraphic classification.

Lithostratigraphic units are based primarily on the lithologic character of rocks—sedimentary, igneous, and metamorphic. The fossil content of lithostratigraphic units may in certain cases be an important distinguishing element in their recognition, not because of the age significance of the fossils, but because of their diagnostic lithologic (physical) characterization. Coquinas, algal reefs, radiolarites, oyster beds, and coal beds are good examples.

Inasmuch as each lithostratigraphic unit was formed during a specific interval of geologic time, it has not only lithologic significance but also chronostratigraphic significance. The concept of time, however, properly plays little part in establishing or identifying lithostratigraphic units and their boundaries. Lithologic character is generally influenced more strongly by conditions of origin than by time of origin; nearly identical rock types are repeated time and again in the stratigraphic sequence, and the boundaries of almost all lithostratigraphic units eventually cut across isochronous surfaces as they are traced laterally.

Biostratigraphic units are based on the fossil content of the rocks. The selection and establishment of biostratigraphic units is not determined by the lithologic composition of the rock strata, except that the presence or absence of fossils, and the kind of fossils present, may be related to the type and facies of the rocks in which they are found.

Lithostratigraphic and biostratigraphic units are fundamentally different kinds of stratigraphic units and are based on different distinguishing criteria. The boundaries of the two may locally coincide, but usually they lie at different stratigraphic horizons or cross each other. Lithostratigraphic and biostratigraphic units differ in another respect; while all sequences of layered rocks—sedimentary, igneous, and metamorphic—can be subdivided into lithostratigraphic units, biostratigraphic units can be recognized only in rocks containing fossil remains.

Both lithostratigraphic and biostratigraphic units rather closely reflect environment of deposition, but biostratigraphic units are much more strongly influenced by, and indicative of, geologic age. They are also less repetitive in character, because they are based to a great extent on the evolutionary changes of plants and animals.

Lithostratigraphy and biostratigraphy are not only invaluable initial steps in working out the stratigraphy of an area; they are important and continuing stratigraphic disciplines in themselves. In many areas they are the fundamental or even the only means of stratigraphic classification. Lithostratigraphic and biostratigraphic units are indispensable objective units, essential in picturing the lithologic constitution and geometry of the Earth's strata and the development of life and past environments on the Earth.

In contrast to lithostratigraphic and biostratigraphic units, which are relatively objective units limited to the actual occurrence of a certain lithology or of certain fossil specimens, *chronostratigraphic* units are defined as encompassing all rocks formed within certain time spans of Earth history regardless of their content. By definition, these units everywhere include rocks of only a certain age and their boundaries are everywhere isochronous. Although lithostratigraphic and biostratigraphic units are largely established and distinguished on the basis of observable physical features, chronostratigraphic units are identified on the basis of their time of formation—a more interpretive character.

Both lithostratigraphic units and biostratigraphic units are invaluable aids to developing a sound chronostratigraphic classification. Because of the widespread distribution of fossils in the Earth's strata, and the irreversibility of biologic evolution, fossils have provided the outstanding guide for dating and time-correlation of Phanerozoic sedimentary rocks. Biostratigraphic units, in particular, frequently approximate chronostratigraphic units, and in practice the two kinds of units can closely correspond. However, although biostratigraphic correlation may approach time-correlation, biostratigraphic units are fundamentally *not* the same as chronostratigraphic units. As illustrated in Figure 14, p. 89, the boundaries of a biostratigraphic zone may diverge from a time horizon for many reasons. Principal among these are changes in depositional facies, variations in conditions for fossilization and preservation of fossils, vagaries of fossil discovery, time required for migration, and geographic differences in evolutionary development. Biostratigraphic units cannot be recognized at all in igneous rocks and in high-grade metamorphic sequences. Even in unaltered sediments there are many poorly fossiliferous or nonfossiliferous intervals. Even so, the contribution of biostratigraphy to chronostratigraphy has been enormous and many of the difficulties in the use of individual biostratigraphic units as time-markers can be solved by the use of several laterally interlocking biozones and biohorizons.

Lithostratigraphic units or lithostratigraphic horizons may also be excellent guides to approximate time-correlation over fairly extensive areas but as in the case of biostratigraphic units these lithostratigraphic units are not the same as chronostratigraphic units because they are not bounded everywhere by isochronous surfaces.

Chronostratigraphic classification, which utilizes information available from all other kinds of stratigraphic classification, stands out as an ultimate goal of stratigraphy. Chronostratigraphic units, as divisions of strata based on geologic time, are in principle worldwide in extent and provide a basis and systematic framework for unraveling the geologic history of the Earth. Chronostratigraphic units, furthermore, are important in providing a worldwide basis for communication and understanding.

The three above-mentioned kinds of stratigraphic units and their corresponding fields of stratigraphic investigation are perhaps the oldest and most commonly used. However, there are many other fruitful lines of stratigraphic endeavor and many other kinds of stratigraphic units which, under appropriate circumstances and for certain objectives, are very useful. Thus we may find it useful to recognize stratigraphic units or horizons based on electric log characters, magnetic reversals, seismic properties, chemical changes, or any of many other characters or properties of rock strata. No one can or need use all the possible kinds of stratigraphic tools or units that are potentially available, but the way should be kept open within the definition and scope of stratigraphy to apply any that give promise of being useful.

Appendix A
Membership of Subcommission on Stratigraphic Classification (1974)

INDIVIDUAL MEMBERS

Almela, A. (Spain)
Alvarez, A. (Mexico)
Barbieri, F. (Italy)
Biq Chingchang (Taiwan)
Bolli, H. M. (Switzerland)
Boškov-Štajner, Z. (Jugoslavia)
Burollet, P. F. (France)
Butterlin, J. (France)
†Chakravarty, S. C. (India)
Chlupáč, I. (Czechoslovakia)
Cohee, G. V. (USA)
Collins, B. W. (New Zealand)
‡Deraniyagala, P. (Sri Lanka)
Drooger, C. W. (Netherlands)
Dubertret, L. (France)
Erben, H. K. (Germany)
Fairbridge, R. (USA)
Fisher, N. H. (Australia)
George, T. N. (UK)
Glaessner, M. F. (Australia)
Gorsky, I. I. (USSR)
Hanzawa, S. (Japan)
Harland, W. B. (UK)
Hedberg, H. D. (USA)
Heide, S. van der (Netherlands)
Henningsmoen, G. (Norway)
Kretzoi, M. (Hungary)
Lawson, J. D. (UK)

Leckwijck, W. P. van (Belgium)
Lepersonne, J. (Belgium)
*Longoria, J. F. (Mexico)
Lüttig, G. (Germany)
Martinsson, A. (Sweden)
Maubeuge, P. L. (France)
McKee, E. D. (USA)
‡Moore, R. C. (USA)
Murray, G. E. (USA)
Nalivkin, D. V. (USSR)
Pomerol, C. (France)
Rankama, K. (Finland)
†Reiss, Z. (Israel)
Renz, H. H. (Switzerland—USA)
Rutsch, R. F. (Switzerland)
Salvador, A. (Venezuela—USA)
Sapunov, I. G. (Bulgaria)
Sastri, V. V. (India)
Sigal, J. (France)
Stainforth, R. M. (Canada)
Stöcklin, J. (Iran)
Størmer, L. (Norway)
Strand, T. (Norway)
Stubblefield, J. (UK)
Toriyama, R. (Japan)
Troelsen, J. C. (Brazil)
Waterhouse, J. B. (Australia)
Zagwijn, W. H. (Netherlands)

*Membership application being processed.
†Retiring.
‡Died 1974.

INDIVIDUAL MEMBERS EX OFFICIO

IUGS Commission on Stratigraphy, D. J. McLaren, Chairman
Subcommission on Cambrian Stratigraphy, A. R. Palmer, Chairman
Subcommission on Carboniferous Stratigraphy, A. Bouroz, Chairman
Subcommission on Cretaceous Stratigraphy, R. Laffitte, Convenor
Subcommission on Geochronology, E. Jäger, Chairman
Subcommission on Gondwana Stratigraphy, E. P. Plumstead, Chairman
Subcommission on Jurassic Stratigraphy, P. L. Maubeuge, Chairman
Subcommission on Neogene Stratigraphy, R. Selli, Chairman
Regional Committee on Northern Neogene Stratigraphy, P. Cambridge, Chairman
Regional Committee on Mediterranean Neogene Stratigraphy, J. Seneš, Chairman
Regional Committee on Pacific Neogene Stratigraphy, N. Ikebe, Convenor
Subcommission on Devonian Stratigraphy, H. K. Erben, Convenor
Subcommission on Magnetic Polarity Time Scale, N. D. Watkins, Convenor
Subcommission on Precambrian Stratigraphy, K. Rankama, President
Subcommission on Quaternary Stratigraphy, V. Sibrava, Chairman
Subcommission on Silurian Stratigraphy, N. Spjeldnaes, Convenor
Subcommission for Stratigraphic Lexicon, C. Lorenz, Chairman
Subcommission on Stratigraphic Classification, H. D. Hedberg, Chairman
Subcommission on Paleogene Stratigraphy, V. V. Menner, Convenor
Subcommission on Triassic Stratigraphy, H. Zapfe, Convenor
Subcommission on Ordovician Stratigraphy, A. Williams, Convenor
Subcommission on Permian Stratigraphy, D. L. Stepanov, Convenor
Regional Committee on European Quaternary Stratigraphy, G. Lüttig, Chairman
Regional Committee on North American Quaternary, R. F. Flint, Chairman
Regional Committee on African Quaternary Stratigraphy, P. Biberson, Chairman
Regional Committee on Stratigraphic Correlation for ECAFE Region, L. Stach, Convenor

Working Group on Upper Cretaceous (Maastrichtian) Stratigraphy, A. A. Thiadens, Chairman

Working Group for Correlation of Cretaceous and Cenozoic Marine Deposits in Marine Geology, H. Bolli, Chairman

Working Group on the Dinantian Subsystem, E. Paproth, Chairman

Working Group on the Jurassic-Cretaceous Boundary, D. C. Patrulius, Convenor

Working Group on the Mississippian-Pennsylvanian Boundary, M. Gordon, Chairman

Working Group on the Namurian Series, W. H. C. Ramsbottom, Chairman

Working Group on the Neogene-Quaternary Boundary, K. V. Nikiforova, Chairman

Working Group on the Precambrian-Cambrian Boundary, J. W. Cowie, Convenor.

ORGANIZATIONAL MEMBERS

AMERICAN COMMISSION on Stratigraphic Nomenclature (CANADA, MEXICO, USA), R. Macqueen, Chairman

(ARGENTINA) Comité Argentino de Nomenclatura Estratigráfica, E. O. Rolleri, President

(AUSTRALIA) Committee on Stratigraphic Nomenclature, E. K. Carter, Representative

(AUSTRIA) Geological Society of Vienna, C. Exner, Representative

(BELGIUM) Geological Council of Belgium, A. Delmer, Secretary

*(BRAZIL) University of São Paulo, Stratigraphic Committee of Instituto de Geociencias

(BURMA) Stratigraphic Committee of Burma, B. T. Haq, Representative CENTRAL AMERICAN Stratigraphic Commission, G. Dengo, Chairman

(COSTA RICA) Geological Survey, L. F. Sandoval M., Director

(CUBA) Instituto de Geologia de la Academia de Ciencias, N. A. Mayo, Director

(EGYPT) Stratigraphic Committee of Egypt, M. N. H. El Gezeery, Representative

(FRANCE) Comité Francais de Stratigraphie, A Blondeau, Representative

(GERMANY) Stratigraphic Commission of Germany, H. Hölder, Chairman

(GHANA) Geological Survey Department, The Acting Director

(GUYANA) Geological Survey of Guyana, S. Singh, Director

(HUNGARY) Stratigraphical Commission, J. Fülöp, President

(INDIA) Committee on the Stratigraphic Nomenclature of India, C. Karunakaran, Chairman and Director General

(INDONESIA) Geological Survey, H. M. S. Hartono, Representative

(INQUA) Working Group to cooperate with ISSC, W. H. Zagwijn, Chairman

(IRELAND) Stratigraphic Standing Committee of Ireland, Irish National Committee for Geology, C. H. Holland, Secretary

(ISRAEL) Geological Survey, Stratigraphic Committee, M. Raab, Chairman

(ITALY) Commisione Stratigrafica del Comitato Geologico Italiano, M. B. Cita, Representative

(JAPAN) Geological Survey, I. Kobayashi, Director

(KOREA) Geological Society of Korea, K. H. Chang, Representative

(MALAYSIA) Geological Society of Malaysia Committee on Stratigraphic Nomenclature, The Acting Director

(NETHERLANDS) Geological Survey, A. A. Thiadens, Director

(NEW ZEALAND) Geological Society, I. G. Speden, Representative

(NEW ZEALAND) Geological Survey, I. G. Speden, Representative

(NORWAY) Commission on Stratigraphy, K. O. Bjørlykke, Representative

*(ROMANIA) Institutul Geologic, D. Patrulius, Representative

(SOUTH AFRICA) Committee for Stratigraphy, L. E. Kent, Chairman

(SWAZILAND) Geological Survey and Mines Department, The Acting Director

(SWEDEN) National Committee of Geology Stratigraphic Committee, A. Martinsson, Chairman

(SYRIA) Working Group for Revision of Stratigraphic Lexicon on Syria, M. Khoury, Representative

(TANZANIA) Mineral Resources Division, Assistant Commissioner for Mineral Resources

(THAILAND) Department of Mineral Resources, K. Pitakpaivan, Representative

(TURKEY) Commission for Stratigraphic Nomenclature, T. Norman, Chairman

(UK) Geological Society of London Stratigraphic Code Sub-committee, N. F. Hughes, Chairman

(USSR) Commission on Stratigraphic Classification, Terminology and Nomenclature, A. I. Zhamoida, President

(VENEZUELA) Comisión Venezolana de Estratigrafía y Ter-minologiá, C. Petzall, Executive Secretary

(SCAR) Working Group on Geology, British Antarctic Survey, I. R. McLeod, Secretary

*Membership application being processed.

CHARTER MEMBERS*

Almela, A. (Spain)	Lotze, F. (Germany)
Cohee, G. V. (USA)	McKee, E. D. (USA)
DeFord, R. K. (USA)	McLaren, D. J. (Canada)
Delépine, G. (France)	Mouta, F. (Portugal)
Dubertret, L. (Lebanon)	Murray, G. E. (USA)
Fisher, N. H. (Australia)	Pamir, H. N. (Turkey)
Glaessner, M. F. (Australia)	Petkovic, K. V. (Yugoslavia)
Harrington, H. J. (Argentina)	Prantl, F. (Czechoslovakia)
Hedberg, H. D. (USA)	Pustovalov, L. V. (USSR)
Henningsmoen, G. (Norway)	Raggatt, H. G. (Australia)
Henson, F. R. S. (Great Britain)	Rutsch, R. F. (Switzerland)
Kegel, W. (Germany-Brazil)	Salvador, A. (Venezuela-USA)
Kellum, L. B. (USA)	Schindewolf, O. H. (Germany)
Khan, N. M. (Pakistan)	Sorgenfrei, T. (Denmark)
Kühn, O. (Australia)	Straelen, V. van (Belgium)
Lecompte, M. (Belgium)	Troelsen, J. C. (Denmark-Brazil)

The names of members who have participated actively in the work on the *Guide* are recorded in the voting lists given in ISSC Reports 1-7 and in the discussion sections of the ISSC Circulars.

*As recorded in ISSC Circular-1 of March 7, 1955.

Appendix B
Published Reports of Subcommission and Library Depositories

ISSC Report 1. Principles of stratigraphic classification and terminology, 1961, *Proc. 21st Int. Geol. Cong. (Norden),* Part 25, 38 p. (Order from 21st Int. Geol. Cong., Øster Voldgade 7, Copenhagen K, Denmark, 5 Danish Kroner per copy.) Translated to Italian, 1963, *Riv. Ital. Pal. Strat.* v. 69, no. 3, p. 429–455. Jugoslav translation by Nada Glumicic-Holland and Zagorka Boškov-Štajner, 1967, *Nafta,* v. 18, nos. 3–4, p. 95–111.

ISSC Report 2. Definition of geologic systems, 1964, *Proc. 22nd Int. Geol. Cong. (India),* Part 18, 26 p. (Reprinted in large part in *Am. Assoc. Petroleum Geol. Bull.,* 1965, v. 49, no. 10, p. 1694–1703.)

ISSC Report 3. Preliminary report on lithostratigraphic units, 1970, *24th Int. Geol. Cong. (Canada),* 30 p. (Order from Secretary-General, 24th Int. Geol. Cong., 601 Booth Street, Ottawa 4, Canada, $1.00 per copy.)

ISSC Report 4. Preliminary report on stratotypes, 1970, *24th Int. Geol. Cong. (Canada),* 39 p. (Order from Secretary-General, 24th Int. Geol. Cong., 601 Booth Street, Ottawa 4, Canada, $1.00 per copy.)

ISSC Report 5. Preliminary report on biostratigraphic units, 1971, *24th Int. Geol. Cong. (Canada),* 50 p. (Order from Secretary-General, 24th Int. Geol. Cong., 601 Booth Street, Ottawa 4, Canada, $1.00 per copy.)

ISSC Report 6. Preliminary report on chronostratigraphic units, 1971, *24th Int. Geol. Cong. (Canada),* 39 p. (Order from Secretary-General, 24th Int. Geol. Cong., 601 Booth Street, Ottawa 4, Canada, $1.00 per copy.)

ISSC Report 7. An international guide to stratigraphic classification, terminology and usage—Introduction and summary, 1972, *Lethaia,* v. 5, no. 3, p. 283–323; *Boreas,* 1972, v. 1, no. 3, p. 199–239. Spanish translation by Cecilia Petzall, 1973, *Bol. Geol.,* v. 11, no. 22, p. 287–331, Caracas, Venezuela.

In addition to the published reports, the Subcommission has issued to its members 46 Circulars (1955–1974) totaling more than 2000 pages. Complete files of these Circulars are available at the following depository libraries, most of which can furnish copies at cost of reproduction:

Geological Survey of Canada—Library
601 Booth Street, Room 350
Ottawa 4, Ontario, Canada K1A OE8

Mineralogisk Museum—Library
Øster Voldgade 5
1350 K, Copenhagen, Denmark

Geological Society of London—Library
Burlington House
London W.1, England, U.K.

British Museum (Natural History)—Paleontology Library
Cromwell Road
London S.W. 7, England, U.K.

Geological Society of France—Library
77, rue Claude Bernard
Paris 5^e, France

Bundesanstalt für Bodenforschung—Library
31 Hannover-Bucholz
Alfred-Bentz-Haus
Postfach 54, Stilleweg 2, Germany

Geological Survey of India—Library
27, Jawaharlal Nehru Road
Calcutta 13, India

Department of Geology—Library
Guyot Hall, Princeton University
Princeton, New Jersey   08540, USA

Texas Tech University Library
P. O. Box 4079
Lubbock, Texas, 79409, USA

U. S. Geological Survey—Library
Reston, Virginia, 22092, USA

All-Union Geological Library
Ministry of Geology of USSR
Sredny prospect 72-B, Leningrad, 199026, USSR

Appendix C
National or Regional
Stratigraphic Codes

1933 **Classification and nomenclature of rocks,** G. H. Ashley et al., Geol. Soc. America Bull., v. 44, p. 423–459; Am. Assoc. Petroleum Geol. Bull., v. 17, no. 7, p. 843–868. Republished in Am. Assoc. Petroleum Geol. Bull., v. 23, no. 7, p. 1068–1088, 1939.

1942 **Rules of geological nomenclature of the Geological Survey of Canada,** Geol. Survey Canada. Republished in Am. Assoc. Petroleum Geol. Bull., v. 32, no. 3, p. 366–367, 1948.

1948 **Stratigraphic nomenclature in Australia,** M. F. Glaessner, H. G. Raggatt, C. Teichert, and D. E. Thomas, Australian Jour. Sci., v. 11, no. 1, p. 7–9.

1950 **Australian code of stratigraphic nomenclature,** Australian Jour. Sci., v. 12, no. 5, p. 170–173.

1952 **Code of stratigraphic nomenclature of the Geological Society of Japan,** Geol. Soc. Japan, Jour., v. 58, p. 112–113. In Japanese with stratigraphic unit terms in English.

1954 **Stratigraficheskie i geokhronologicheskie podrazdeleniya** (Stratigraphic and geochronologic subdivisions—their principles, contents, terminology, and rules of use) (ed. L. S. Librovich), VSEGEI, Gosgeoltehizdat, Moscow, 85 p.

1956 **Australian code of stratigraphic nomenclature (2nd ed.),** Australian Jour. Sci., v. 18, no. 4, p. 117–121.

1956 **Stratigraficheskaya klassifikatsiya i terminologiya** (Stratigraphic classification and terminology), Report by Interdepartmental Stratigraphic Committee of USSR (ed. A. P. Rotay), Gosgeoltekhizdat, Moscow. English translation in Int. Geol. Rev., v. 1, no. 2, p. 22–38, 1959.

1959 **Australian code of stratigraphic nomenclature (3rd ed.),** Jour. Geol. Soc. Australia, v. 6, pt. 1, 63–70.

1960 **Chinese code of stratigraphic nomenclature** (in Chinese), Geol. Soc. China Proc., no. 3, p. 2–5. (Fide Biq Chingchang).

1960 **Ceskoslovenská stratigrafická terminologie** (Czechoslovak stratigraphic terminology), Vestnik UUG, v. 35, p. 95–110, Prague.

1960 **Stratigraficheskaya klassifikatsiya i terminologia** (Stratigraphic classification and terminology), Interdepartmental Stratigraphic Committee of the USSR, 2nd revised edition (ed. A. P. Rotay), (in Russian and English), Gosgeoltehizdat, Moscow, 60 p.

1961 **Code of stratigraphic nomenclature,** American Commission on Stratigraphic Nomenclature, Am. Assoc. Petroleum Geol. Bull., v. 45, no. 5, p. 645–665.

1961 **Código de nomenclatura estratigráfica** (Code of stratigraphic nomenclature), Spanish translation of the Code of Stratigraphic Nomenclature of the American Commission on Stratigraphic Nomenclature, prepared by Manuel Alvarez, Jr., Mexico, D.F., 28 p.

1961 **Regler for Norsk stratigrafisk nomenklatur** (Code of stratigraphic nomenclature for Norway), Commission on Stratigraphy of Norway, Norges Geologiske Undersøkelse, no. 213, p. 224–233. (in Norwegian and English).

1962 **Principes de classification et de nomenclature stratigraphiques** (Principles of stratigraphic classification and nomenclature), Comité Francais de Stratigraphie, 15 p., available from A. Blondeau, Géologie des Bassins Sedimentaires, Université Paris VI, 4, place Jussieu, 75230 Paris Cédex 05.

1962 **Stratigraphic code of Pakistan,** Stratigraphic Nomenclature Committee of Pakistan, Mem. Geol. Surv. Pakistan, v. 4, pt. 1, p. 1–8.

1962 **Codice di nomenclatura stratigrafica secondo i Nord-Americani,** Riv. Ital. Pal. Strat., v. 68, no. 1, p. 115–148.

1964 **Australian code of stratigraphic nomenclature (4th edition),** Committee on Stratigraphic Nomenclature of Geological Society of Australia, Jour. Geol. Soc. Australia, v. 11, pt. 1, p. 165–171; pt. 2, p. 342. (Reprinted in 1973 with corrigenda and additional notes.)

1965 **Stratigraficheskaya klassifikatsiya, terminologia i nomenklatura** (Stratigraphic classification, terminology and nomenclature), (ed. A. I. Zhamoida), Izdatelstvo Nedra, Leningrad, 70 p., English translation in Int. Geol. Rev. v. 8, no. 10, pt. 2, p. 1–36, 1966.

1965 **Project of a stratigraphic code** (in Chinese), People's Republic of China Stratigraphic Conference, Pekin, 54 p. (Fide Zhamoida).

1966 **Litostratigrafski yedinitsi—shchnost, nomenklatura, i klassifikatsiya** (Lithostratigraphical units—nature, nomenclature and classification), prepared by T. Nikolov, I. Sapunov, J. Stephanov, Y. Tenchov, and Kh. Khrischev, for Bulgarian Geological Society, in Bulgarian, with English summary, Rev. Bulgarian Geol. Soc., v. 27, pt. 3, p. 233–247.

1967 **Report of Stratigraphical Code Subcommittee of Geological Society of London,** Proc. Geol. Soc. London, no. 1638, p. 75–87.

1967 **Guide to stratigraphic nomenclature,** New Zealand Geol. Soc., 20 p.

1968 **Malaysian code of stratigraphical nomenclature,** Geol. Soc. of Malaysia, Kuala Lumpur, 11 p. (reprinted).

1968 **International Geological Correlation Program, United Kingdom Contribution,** British National Committee for Geology, Royal Society, 43 p. (Contains revised edition of 1967 Report of the Stratigraphical Code Sub-Committee.)

1968 **Stratigrafi siniflama ve adlama kurallari** (Turkish code of stratigraphic nomenclature), Turkish Stratigraphic Committee, 28 p., Ankara, Turkey.

1968 **Preliminarni stratigrafski kodeks** (Preliminary stratigraphic code), Zagorka Boskov-Stajner, Nafta v. 19, no. 12 (Geologija i geofizika) December 1968, p. 529–534, Zagreb, Yugoslavia. (English translation by D. Z. Briggs made for U. S. Geological Survey, 1970, 17 p.)

1969 **Codice Italiano di nomenclatura stratigrafica** (Italian code of stratigraphic nomenclature), prepared by A. Azzaroli and M. Bianca Cita, with collaboration of R. Selli, Servizio Geologico d'Italia Boll., v. 89, (1968), p. 3–22.

1969 **Recommendations on stratigraphical usage,** Geol. Soc. London Proc., no. 1656, p. 139–166. (2nd revision of 1967 Report of the Stratigraphic Code Sub-Committee).

1969 **Key to the interpretation and nomenclature of Quaternary stratigraphy,** compiled by G. W. Lüttig, R. Papep, R. G. West, and W. H. Zagwijn, INQUA Commission on Stratigraphy. First and provisional edition 46 p., Hannover, Germany.

1970 **Code of stratigraphic nomenclature (2nd edition),** American Commission on Stratigraphic Nomenclature, 22 p., published by Am. Assoc. Petroleum Geol., Tulsa, Oklahoma.

1970 **Codigo de nomenclatura estratigráfica (Segunda edición)** (Code of stratigraphic nomenclature, 2nd edition), Spanish translation of the second edition of the Code of Stratigraphic Nomenclature of the American Commission on Stratigraphic Nomenclature, prepared by D. A. Córdoba and Z. de Cserna, 28 p., Mexico, D.F.

1970 **Proekt stratigraficheskogo kodeksa SSSR** (Project of a stratigraphic code for USSR), compiled by A. I. Zhamoida, O. P. Kovalevskiï, A. I. Moisseeva, and V. I. Yarkin (Departmental Stratigraphic Committee of USSR), 55 p.

1971 **South African code of stratigraphic terminology and nomenclature,** Geol. Soc. South Africa, Trans., v. 74, p. 111–131.

1971 **Code of stratigraphic nomenclature of India,** India Geol. Survey Misc. Publ. no. 20, 28 p.

1972 **A concise guide to stratigraphical procedure,** W. B. Harland, et al. (for Stratigraphy Committee of Geol. Soc. London), Geol. Soc. London Quart. Jour., v. 128, p. 295–305.

1972 **Main principles of the draft of the USSR Stratigraphic Code,** A. I. Zhamoida, O. P. Kovalevskii, V. V. Menner, A. I. Moisseeva, and V. I. Yarkin, Report at the meeting of the International Subcommission on Stratigraphic Classification, Montreal, 1972. Published in Leningrad, 1972, in Russian and English, 14 p.

1973 **Empfehlungen zur Handhabung der stratigraphischen, insbesondere lithostratigraphischen Nomenclatur in der Schweiz** (Recommendations on the application of stratigraphic, especially lithostratigraphic, nomenclature in Switzerland), Arbeitsgruppe für Stratigraphische Terminologie, Schweizerische Geologische Komission, Eclog. Geol. Helv., v. 66, no. 2, p. 479–492.

1973 **Sandi stratigrafi Indonesia,** Komisi Sandi Stratigrafi Indonesia, 19 p.

1975 **Sandi stratigrafi Indonesia** (Stratigraphic Code of Indonesia) (Revised edition), Commission for Stratigraphic Code of Indonesia, 19 p. (Published in Indonesian and in English).

Appendix D
Vote of Members
of the Subcommission
on Publication
of the International
Stratigraphic Guide

Individual Members	Country	In Favor of Publication
Almela	Spain	Yes
Alvarez	Mexico	Yes
Barbieri	Italy	Yes
Biq Chingchang	Taiwan	Yes
Bolli	Switzerland	Yes
Boškov-Štajner	Jugoslavia	Yes
Burollet	France	Yes
Butterlin	France	Yes
Chlupáč	Czechoslovakia	Yes
Cohee	USA	Yes
Collins	New Zealand	Yes
Drooger	Netherlands	Yes
Dubertret	France	Yes
Erben	Germany (F.R.) (Vote recorded in *Ex Officio* List)	
Fairbridge	USA	Yes
Fisher	Australia	Yes
George	UK	Yes
Glaessner	Australia	Yes
Hanzawa	Japan	Yes
Harland	UK	Yes
Hedberg	USA	Yes
Heide, van der	Netherlands	Yes
Henningsmoen	Norway	Yes
Lawson	UK	Yes
Leckwijck, van	Belgium	Yes

Lepersonne	Belgium	Yes
Martinsson	Sweden	Yes
Maubeuge	France (Vote recorded in *Ex Officio* list)	
McKee	USA	Yes
Murray	USA	Yes
Pomerol	France	Yes
Rankama	Finland	Yes
Reiss	Israel	Yes
Renz	Switzerland—USA	Yes
Rutsch	Switzerland	Yes
Salvador	Venezuela—USA	Yes
Sapunov	Bulgaria	Yes
Sastri	India	Yes
Sigal	France	Yes
Stainforth	Canada	Yes
Stöcklin	Iran	Yes
Størmer	Norway	Yes
Strand	Norway	Yes
Stubblefield	UK	Yes
Toriyama	Japan	Yes
Troelsen	Brazil	Yes
Waterhouse	Australia	Yes
Zagwijn	Netherlands	Yes

Individual Members ex officio	In Favor of Publication
IUGS Commission on Stratigraphy (McLaren)	Yes
Subcommission on Cambrian Stratigraphy (Palmer)	Yes
Subcommission on Carboniferous Stratigraphy (Bouroz)	Yes
Subcommission on Devonian Stratigraphy (Erben)	Yes (provisional)
Subcommission on Geochronology (Jäger)	Yes
Subcommission on Jurassic Stratigraphy (Maubeuge)	Yes
Subcommission on Paleogene Stratigraphy (Menner)	No
Subcommission on Silurian Stratigraphy (Spjeldnaes)	Yes
Regional Committee on Mediterranean Neogene Stratigraphy (Seneš)	Yes

Regional Committee on Pacific Neogene Stratigraphy (Ikebe) Yes

Working Group on Dinantian Subsystem (Paproth) Yes

Working Group on Precambrian-Cambrian
Boundary (Cowie) Yes

Organizational Members

AMERICAN COMMISSION on Stratigraphic
Nomenclature (By Oriel, 1974; By Macqueen,
1975) Yes

(ARGENTINA) Comité Argentino de
Nomenclatura Estratigráfica (By Rolleri) Yes

(AUSTRALIA) Stratigraphic Nomenclature
Committee, Geological Society of Australia
(By Carter) Yes

(AUSTRIA) Geological Society of Vienna (By Exner) Yes

(BELGIUM) Comité national belge de Geologie
(By Delmer) Yes

CENTRAL AMERICAN Stratigraphic Commission (By Dengo) Yes

(COSTA RICA) Geological Survey (By Brenes
Monge) Yes

(FRANCE) Comité Francais de Stratigraphie (By
Blondeau) Yes

(GERMANY, FEDERAL REPUBLIC OF) Stratigraphic Commission (By Hölder) No

(HUNGARY) Stratigraphical Commission (By
Fülöp) Yes

(INDONESIA) Geological Survey (By Hartono) Yes

(IRELAND) Stratigraphic Standing Committee
(By Holland) Yes

(ISRAEL) Geological Survey of Israel, Stratigraphic Committee (By Raab) Yes

(ITALY) Comm. Strat. del Com. Geol. Italiano
(By Cita) Yes

(JAPAN) Geological Survey of Japan (By
Kobayashi) Yes

(KOREA) Geological Society of Korea (By Chang)   Yes

(MALAYSIA) Geological Society of Malaysia (By
   Yancey)   Yes

(NETHERLANDS) Geological Survey (By van der
   Heide)   Yes

(NEW ZEALAND) Geological Society (By Speden)   Yes

(NEW ZEALAND) Geological Survey (By Speden)   Yes

(NORWAY)   Commission   on   Stratigraphy,
   Norwegian Geological Survey (By Bjørlykke)   Yes

(ROMANIA) Institutul Geologic (By Patrulius)   Yes

(SOUTH AFRICA) Committee for Stratigraphy
   (By Kent)   Yes

(SWEDEN) National Committee for Geology
   (By Martinsson)   Yes

(THAILAND) Department of Mineral Resources
   (By Pitakpaivan)   Yes

(TURKEY)   Commission   for   Stratigraphic
   Nomenclature (By Norman)   Yes

(UK) Geological Society of London, Stratigraphic
   Code Subcommittee (By Hughes)   Yes?

(USSR) Comm. on Strat. Classif., Term., and
   Nomenclature (By Zhamoida)   No

(VENEZUELA) Com. Venezolana de Estrat. y
   Term. (By Petzall)   Yes

SCAR Working Group on Geology (By McLeod)   Yes

Bibliography of Stratigraphic Classification and Terminology

A comprehensive bibliography of published works on stratigraphic classification, terminology, and procedure is obviously an essential adjunct to the Subcommission's *International Stratigraphic Guide*. It serves to show the background of ideas from which the *Guide* has grown, the evolution of thinking on stratigraphic classification and terminology, and the current status of views on these subjects in various parts of the world. Particularly important, it gives representation to individual or national expressions which may agree or be at variance in certain respects with those contained in the *Guide* and thus gives the user of the *Guide* a broader critical background conducive to eventual improvement of the present edition.

The discriminative selection of items for the bibliography has been a difficult task because the subject matter grades indistinguishably into many other fields and because passages pertinent to the scope of the *Guide* may be buried in publications whose major topic and title show little apparent relation to those of the *Guide*. Similarly the distribution of the literature among so many fields, so many countries, so many languages, and over so long a time span has made the task difficult. The help of the entire membership has been utilized in suggesting titles and checking references. Special recognition is due Professor A. I. Zhamoida for his assistance in supplying references to Russian and Chinese literature.

In spite of all efforts it is certain that some pertinent publications have been overlooked, and it is equally certain that some have been inappropriately included. Criticism and suggestions for amendment will be appreciated. To keep the bibliography to a reasonable length and a reasonably sharp focus, the aim has been to limit it largely to *include* only items falling under the following topics:

1. Principles of stratigraphy, and principles and procedures of stratigraphic classification.

2. Principles and procedures of stratigraphic correlation.

3. Stratigraphic terminology and rules of stratigraphic nomenclature.

4. Initial or early descriptions of certain stratigraphic methods or procedures pertinent to stratigraphic classification.

5. A few particularly instructive specific examples of the application of principles and procedures of stratigraphic classification and correlation.

6. Some outstanding early historical contributions to the development of concepts of stratigraphic classification, terminology, and nomenclature.

7.   Stratigraphic guides, codes, and rules.

8.   Records of meetings containing discussions of stratigraphic terminology and principles of stratigraphic classification.

9.   *Reviews* of pertinent publications bearing on stratigraphic terminology and principles of stratigraphic classification.

10.   Reports and Circulars of the International Subcommission on Stratigraphic Classification (ISSC).

Because of the same considerations of keeping to a practical size and scope, it has been necessary to *exclude* from the bibliography the vast number of excellent works dealing entirely with general descriptive or interpretative stratigraphy, many of which at first thought might have seemed appropriate for the bibliography. Among these are the following kinds of papers:

1.   Studies of the stratigraphy of *specific* geographic areas or specific stratigraphic intervals—unless particularly instructive as to principle.

2.   Definitions of *specific* stratigraphic units—unless particularly useful as models.

3.   Application of methods of correlation or stratigraphic study—unless setting forth new concepts or techniques closely concerned with stratigraphic classification.

4.   Lexicons of named stratigraphic units.

5.   *Specific cases* of stratigraphic zonation, correlation, or boundary definition by paleontological, isotopic, and other methods—unless involving new contributions to principles or terminology.

6.   Records or minutes of meetings of stratigraphic committees or groups that do not include information pertinent to general stratigraphic classification or stratigraphic terminology.

Some variations in bibliographic form may be noticed. However, the principal objective has been to provide a *usable* citation—one which is accurate in name of author, date, title of paper, name of publisher, volume number, and page number—even though occasionally lacking in consistency of style.

The editor wishes to express his grateful appreciation to all members and others who have assisted in supplying and checking references and to the staffs of numerous libraries who have been extremely helpful.

*   An asterisk preceding a reference indicates that it is a Russian reference supplied through the courtesy of Professor A. I. Zhamoida. Also, a list of papers on stratigraphic classification published in the People's Republic of China in recent years is reproduced at the end of the Bibliography from Zhamoida, Kovalevskii, and Moiseeva (1969).

Adams, C. G., 1965, The foraminifera and stratigraphy of the Melinau Limestone, Sarawak, and its importance in Tertiary correlation: Geol. Soc. London Quart. Jour., v. 121, p. 283–338.

————, 1970, A reconsideration of the East Indian letter classification of the Tertiary: British Museum (Natural History) Bull., Geology, v. 19, no. 3, p. 87–137.

Adams, J. A. S., and **J. J. W. Rogers,** 1961, Bentonites as absolute time-stratigraphic calibration points: *in* Geochronology of rock systems, N. Y. Acad. Sci. Ann., v. 91, Art. 2, p. 390–396.

Adegoke, O. S., 1970, Principles of stratigraphy: *in* Stratigraphy: an interdisciplinary symposium (ed. Daniels and Freeth), Ibadan Univ., Inst. African Studies, Occasional Publications no. 19, p. 5–15.

Ager, D. V., 1963, Jurassic stages: Nature, v. 198, no. 4885, p. 1045–1046.

————, 1964, The British Mesozoic Committee: Nature, v. 203, no. 4949, p. 1059.

————, 1964, The Luxembourg Colloquium: Geol. Mag., v. 101, p. 471–472.

————, 1967, Bases as a basis of Upper Jurassic correlation: Manuscript presented at International Symposium on Upper Jurassic Stratigraphy in USSR, June 1967, and to be published in Russian, 9 p.

————, 1970, The Triassic system in Britain and its stratigraphical nomenclature: Geol. Soc. London Quart. Jour., v. 126, nos. 501/2, parts 1 and 2, p. 3–17.

————, 1973, The nature of the stratigraphical record: Wiley, New York, 114 p.

Agnew, A. F., 1957, Discussion of Note 17—Suppression of homonymous and obsolete stratigraphic names, American Commission on Stratigraphic Nomenclature: Am. Assoc. Petroleum Geol. Bull., v. 41, no. 8, p. 1889–1890.

Albear, J. F. de, et al., 1968, Las formaciones geologicas y su importancia en la solución de algunos problemas geologicos: Acad. Ciencias, Cuba, Inst. Geol., Ser. Geol., no. 2, 16 p.

Alberti, F. von, 1834, Beitrag zu einer Monographie des Bunten Sandsteins, Muschelkalks und Keupers und die Verbindung dieser Gebilde zu einer Formation: Stuttgart und Tübingen, 366 p. (see p. 1–16 and 300–343).

Alcock, F. J., 1934, Report of the National Committee on Stratigraphical Nomenclature: Royal Soc. Canada, Trans., ser. 3, v. 28, sec. 4, p. 113–121.

Alimov, A. I., 1970, Opredeleniye ponyatiya "regional noya stratigraficheskoye polrazdeleniye" (Definition of the concept "regional stratigraphic subdivision"): Sov. Geol., no. 12, p. 108–113.

Allan, R. S., 1934, On the system and stage names applied to subdivisions of the Tertiary strata in New Zealand: New Zealand Inst. (Royal Soc. of New Zealand) Trans. and Proc., v. 63, p. 81–108.

————, 1948, Geological correlation and paleoecology: Geol. Soc. America Bull., v. 59, no. 1, p. 1–10.

————, 1956, Report of Chairman: Standing committee on datum-planes in the geological history of the Pacific region: 8th Pacific Sci. Cong. (1953), Proc., v. 2, p. 325–423; Canterbury Univ. College, Christchurch, New Zealand, 83 p.

————, 1966, The unity of stratigraphy: New Zealand Jour. Geol. Geophys., v. 9, no. 4, p. 491–494.

Allasinaz, A., 1964, Sulla nomenclatura stratigrafica del Carnico: Riv. Ital. Paleont. Strat., v. 70, no. 1, p. 3–14.

Allemann, F., et al., 1971, Standard Calpionellid zonation (Upper Tithonian-Valanginian) of the western Mediterranean Province: *in* 2nd Planktonic Conf. (Rome, 1970) Proc., (ed. by A. Farinacci), Edizioni Tecnoscienza, p. 1337–1340.

Allen, P. M., and **A. J. Reedman,** 1968, Stratigraphic classification in Pre-Cambrian rocks: Geol. Mag., v. 105, no. 3, p. 290–297.

Alpern, B., 1970, Le concept de biozone en palynologie houillière: Paläont. Abh., Abt. B, Bd. 3, H. 3/4, p. 277–278.

———, 1970, Notes sur les concepts d'espèce et de biozone: *in* Colloque sur la stratigraphie du Carbonifère, Liège Univ., Cong. Colloq., v. 55, p. 81–89.

——— and **J. J. Liabeuf,** 1969, Palynological considerations on the Westphalian and the Stephanian: Proposition for a parastratotype: 6th Int. Cong. on the stratigraphy and geology of the Carboniferous (Sheffield, 1967), v. 1, p. 109–114.

——— and **S. Durand,** 1972, Les méthodes de la palynologie stratigraphique: *in* Colloque sur les méthodes et tendances de la stratigraphie, Orsay (1970), BRGM France, Mem. 77, pt. 1, p. 201–216.

Alvarez, M., Jr., 1957, Comments on Report 5 (of Am. Comm. Strat. Nomen.)—Nature, usage and nomenclature of biostratigraphic units: Am. Assoc. Petroleum Geol. Bull., v. 41, no. 8, p. 1888–1889.

———, 1959, Versión Castellana de la redacción preliminar del Código Estratigráfico: Soc. Geol. Mexicana Bol., v. 22, no. 1, 32 p.

American Commission on Stratigraphic Nomenclature (prepared by **R. C. Moore**), 1947, Note 1—Organization and objectives of the Stratigraphic Commission: Am. Assoc. Petroleum Geol. Bull., v. 31, no. 3, p. 513–518.

——— (prepared by **R. C. Moore**), 1947, Note 2—Nature and classes of stratigraphic units: Am. Assoc. Petroleum Geol. Bull., v. 31, no. 3, p. 519–528.

——— (prepared by **R. C. Moore**), 1948, Note 3—Rules of geological nomenclature of the Geological Survey of Canada: Am. Assoc. Petroleum Geol. Bull., v. 32, no. 3, p. 366–367.

——— (prepared by **W. V. Jones** and **R. C. Moore**), 1948, Note 4—Naming of subsurface stratigraphic units: Am. Assoc. Petroleum Geol. Bull., v. 32, no. 3, p. 367–371.

——— (prepared by **R. F. Flint** and **R. C. Moore**), 1948, Note 5—Definition and adoption of the terms stage and age: Am. Assoc. Petroleum Geol. Bull., v. 32, no. 3, p. 372–376.

——— (prepared by **R. C. Moore**), 1948, Note 6—Discussion of nature and classes of stratigraphic units: Am. Assoc. Petroleum Geol. Bull., v. 32, no. 3, p. 376–381.

——— (prepared by **R. C. Moore**), 1949, Note 7—Records of the Stratigraphic Commission for 1947–1948: Am. Assoc. Petroleum Geol. Bull., v. 33, no. 7, p. 1271–1273.

——— (prepared by **R. C. Moore**), 1949, Note 8—Australian code of stratigraphic nomenclature: Am. Assoc. Petroleum Geol. Bull., v. 33, no. 7, p. 1273–1276.

——— (prepared by **R. C. Moore**), 1949, Note 9—The Pliocene-Pleistocene boundary: Am. Assoc. Petroleum Geol. Bull., v. 33, no. 7, p. 1276–1280.

———, 1949, Report 1—Declaration on naming of subsurface stratigraphic units: Am. Assoc. Petroleum Geol. Bull., v. 33, no. 7, p. 1280–1282.

——— (prepared by **R. C. Moore**), 1950, Note 10—Should additional categories of stratigraphic units be recognized?: Am. Assoc. Petroleum Geol. Bull., v. 34, no. 12, p. 2360–2361.

—————— (prepared by **R. C. Moore**), 1951, Note 11—Records of the Stratigraphic Commission for 1949–1950: Am. Assoc. Petroleum Geol. Bull., v. 35, no. 5, p. 1074–1076.

—————— (prepared by **R. C. Moore**), 1951, Note 12—Divisions of rocks and time: Am. Assoc. Petroleum Geol. Bull., v. 35, no. 5, p. 1076.

—————— (prepared by **H. D. Hedberg**), 1952, Report 2—Nature, usage and nomenclature of time-stratigraphic and geologic-time units: Am. Assoc. Petroleum Geol. Bull., v. 36, no. 8, p. 1627–1638.

——————, 1952, Note 14—Official report of round table conference on stratigraphic nomenclature at Third Congress of Carboniferous Stratigraphy and Geology, Heerlen, Netherlands, June 26–28, 1951: Am. Assoc. Petroleum Geol. Bull., v. 36, no. 10, p. 2044–2048.

—————— (prepared by **R. D. Hutchinson**), 1953, Note 15—Records of the Stratigraphic Commission for 1951–1952: Am. Assoc. Petroleum Geol. Bull., v. 37, no. 5, p. 1078–1080.

—————— (prepared by **J. M. Harrison**), 1955, Report 3—Nature, usage, and nomenclature of time-stratigraphic and geologic-time units as applied to the Precambrian: Am. Assoc. Petroleum Geol. Bull., v. 39, no. 9, p. 1859–1861.

—————— (prepared by **D. J. McLaren**), 1955, Note 16—Records for the Stratigraphic Commission for 1953–1954: Am. Assoc. Petroleum Geol. Bull., v. 39, no. 9, p. 1861–1863.

—————— (prepared by **G. V. Cohee, R. K. DeFord, J. M. Harrison, G. E. Murray,** and **C. H. Stockwell**), 1956, Report 4—Nature, usage, and nomenclature of rock-stratigraphic units: Am. Assoc. Petroleum Geol. Bull., v. 40, no. 8, p. 2003–2014.

—————— (prepared by **R. C. Moore** and **G. V. Cohee**), 1956, Note 17—Suppression of homonymous and obsolete stratigraphic names: Am. Assoc. Petroleum Geol. Bull., v. 40, no. 12, p. 2953–2954.

—————— (prepared by **J. Gilluly**), 1957, Note 18—Records of the Stratigraphic Commission for 1955–1956: Am. Assoc. Petroleum Geol. Bull., v. 41, no. 1, p. 130–133.

—————— (prepared by **H. D. Hedberg, M. Gordon, Jr., E. T. Tozer, H. E. Wood, II,** and **K. Lohman**), 1957, Report 5—Nature, usage, and nomenclature of biostratigraphic units: Am. Assoc. Petroleum Geol. Bull., v. 41, no. 8, p. 1877–1889.

—————— (prepared by **E. D. McKee**), 1957, Discussion of Note 17—Suppression of homonymous and obsolete stratigraphic names: Am. Assoc. Petroleum Geol. Bull., v. 41, no. 8, p. 1889–1891.

—————— (prepared by **J. C. Frye**), 1958, Note 21—Preparation of new stratigraphic code by American Commission on Stratigraphic Nomenclature: Am. Assoc. Petroleum Geol. Bull., v. 42, no. 8, p. 1984–1986.

—————— (prepared by **G. M. Richmond**), 1959, Report 6—Application of stratigraphic classification and nomenclature to the Quaternary: Am. Assoc. Petroleum Geol. Bull., v. 43, no. 3, p. 663–673.

—————— (prepared by **J. C. Frye**), 1959, Discussion of Report 6—Application of stratigraphic classification and nomenclature to the Quaternary: Am. Assoc. Petroleum Geol. Bull., v. 43, no. 3, p. 674–675.

—————— (prepared by **K. E. Lohman**), 1959, Note 22—Records of the Stratigraphic Commission for 1957–1958: Am. Assoc. Petroleum Geol. Bull., v. 43, no. 8, p. 1967–1971.

——————, 1961, Code of stratigraphic nomenclature: Am. Assoc. Petroleum Geol. Bull., v. 45, no. 5, p. 645–665. (Spanish transl. by **M. Alvarez, Jr.,** 1961, 28 p., Editorial Stylo, Mexico, D. F.) (Italian transl., 1962, Riv. Ital. Pal. Strat., v. 68, no. 1, p. 115–148.)

—— (prepared by **G. E. Murray**), 1961, Note 26—Records of the Stratigraphic Commission for 1959–1960: Am. Assoc. Petroleum Geol. Bull., v. 45, no. 5, p. 670–673.

—— (prepared by **J. B. Patton**), 1963, Note 29—Records of the Stratigraphic Commission for 1961–1962: Am. Assoc. Petroleum Geol. Bull., v. 47, no. 11, p. 1987–1991.

——, 1964, Correction to Note 30: Am. Assoc. Petroleum Geol. Bull., v. 48, no. 7, p. 1196.

—— (prepared by **G. V. Cohee**), 1965, Note 31—Records of the Stratigraphic Commission for 1963–1964: Am. Assoc. Petroleum Geol. Bull., v. 49, no. 3, p. 296–300.

——, 1965, Note 32—"Definition of Geologic Systems" by International Subcommission on Stratigraphic Terminology: Am. Assoc. Petroleum Geol. Bull., v. 49, no. 10, p. 1694–1703.

—— (prepared by **P. Harker**), 1967, Note 34—Records of the Stratigraphic Commission for 1964–1966: Am. Assoc. Petroleum Geol. Bull., v. 51, no. 9, p. 1862–1868.

—— (prepared by **F. E. Kottlowski**), 1969, Note 37—Records of the Stratigraphic Commission for 1966–1968: Am. Assoc. Petroleum Geol. Bull., v. 53, no. 10, p. 2179–2186.

——, 1970, Code of stratigraphic nomenclature (2nd ed.): Am. Assoc. Petroleum Geologists, 21 p. (Spanish translation by **D. A. Córdoba** and **Z. de Cserna** (1970) Mexico, D. F., 28 p.)

—— (prepared by **J. A. Wilson**), 1971, Note 39—Records of the Stratigraphic Commission for 1968–1970: Am. Assoc. Petroleum Geol. Bull., v. 55, no. 10, p. 1866–1872.

Andrews, J., and **K. J. Hsu**, 1970, Note 38 (of Am. Comm. Strat. Nomen.)—A recommendation to the American Commission on Stratigraphic Nomenclature concerning nomenclatural problems of submarine formations: Am. Assoc. Petroleum Geol. Bull., v. 54, no. 9, p. 1746–1747.

* **Andrusov, D.,** 1963, Poznámky k stratigrafickému názvoslovin: Geol. Sbornik, v. 14, č 2, p. 319–320.

* —— and **E. Scheibner**, 1964, Návrh slovenskij stratigrafickej klasifikacie a terminologie: Geol. Sbornik, v. 15, č 1, p. 167–172.

Anon., 1970, Informes preliminares sobre "unidades litoestratigraficas" y "estratotipos": Bol. Geologia (Venezuela), v. 11, no. 21, p. 337–396. (Spanish translation of ISSC Circulars 26 and 27).

Anthony, J. W., 1955, Geological stratigraphy: Geochronology, Arizona Univ. Physical Science Bull., no. 2, p. 82–86.

Arduino, G., 1759 or 1760, Letters of Giovanni Arduino to Antonio Vallisnieri, dated Jan. 30, 1759 and March 30, 1759, published in Nuovo raccolta di opuscoli scientifici e filologici del padre abate Angiolo Calogiera, v. 6, p. 99–180.

Arkell, W. J., 1933, The Jurassic System in Great Britain: Clarendon Press, Oxford, 681 p. (see p. 1–37).

——, 1946, Standard of the European Jurassic: Geol. Soc. America Bull., v. 57, p. 1–34.

——, 1951, Review of "Grundlagen und Methoden der Paläeontologischen Chronologie" (3rd ed.) by O.H. Schindewolf: Geol. Mag., v. 88, p. 303–304.

——, 1956, Comments on stratigraphic procedure and terminology: Am. Jour. Sci., v. 254, p. 457–467.

——, 1956, Jurassic geology of the world: Oliver and Boyd, London, 806 p. (see Chapter 1, Classification and correlation, p. 3–14.)

——, 1958, Further comments on stratal terms: Discussion (of Schindewolf comments on stratigraphic terms 1957: Am. Jour. Sci., v. 255, p. 394–399): Am. Jour. Sci., v. 256, no. 5, p. 365.

Arkin, I., M. Braun, and **A. Starinsky,** 1965, Lithostratigraphy, type sections of Cretaceous formations in the Jerusalem-Bet Shemesh Area: Israel Geol. Survey, Stratigraphic Sections, Pub. no. 1, June 1965, p. 2–24.

Arnold, H., 1966, Grundsätzliche Schwierigkeiten bei der biostratigraphischen Deutung phyletischer Reihen: Senckenbergiana Lethaea, v. 47, nos. 5–6, p. 537–547, Frankfurt.

Ashley, G. H., 1932, Stratigraphic nomenclature: Geol. Soc. America Bull., v. 43, no. 2, p. 469–476.

————, 1932, Geologic time and the rock records: Geol. Soc. America Bull., v. 43, no. 2, p. 477–486.

————, 1938, The Canadian System: Topographic and geologic survey progress report 119, Pennsylvania Geol. Survey, 7 p.

———— **et al.,** 1933, Classification and nomenclature of rock units: Geol. Soc. America Bull., v. 44, no. 2, p. 423–459.

———— **et al.,** 1933, Classification and nomenclature of rock units: Am. Assoc. Petroleum Geol. Bull., v. 17, no. 7, p. 843–868.

———— **et al.,** 1939, Classification and nomenclature of rock-units: Am. Assoc. Petroleum Geol. Bull., v. 23, no. 7, p. 1068–1088.

Austin, R., et al., 1970, Les couches de passage du Dévonien au Carbonifère de Hook Head (Irlande) au Bohlen (D.D.R.): Colloque sur la Stratigraphie du Carbonifère, Cong. et Colloques de l'univ. de Liège, v. 55, p. 167–177.

Australian Code of Stratigraphic Nomenclature, 1950 (1st ed.): Australian Jour. Sci., v. 12, no. 5, p. 170–173. (Also see Glaessner et al., 1948).

Australian Code of Stratigraphic Nomenclature, 1956 (2nd ed.): Australian Jour. Sci., v. 18, no. 4, p. 117–121.

Australian Code of Stratigraphic Nomenclature, 1959 (3rd ed.): Geol. Soc. Australia Jour., v. 6, pt. 1, p. 63–70.

Australian Code of Stratigraphic Nomenclature, 1964 (4th ed.): Committee on Stratigraphic Nomenclature of Geol. Soc. of Australia, Geol. Soc. Australia Jour., v. 11, pt. 1, p. 165–171; pt. 2, p. 342. (Reprinted in 1973 with corrigenda and additional notes.)

Azzaroli, A., M. B. Cita, and **R. Selli** (for Com. Strat. del Com. Geol. Ital.), 1968, Codice Italiano di nomenclatura stratigrafica: Boll. Serv. Geol. Italia, v. 89, 1969, Nuova Tecnica Grafica, p. 3–22.

———— and **M. B. Cita,** 1963 (?), Geologia stratigrafica: v. 1, Milano, 262 p. (see p. 1–120).

Barbieri, F., 1971, Comments on some Pliocene stages and on the taxonomy of a few species of *Globorotalia*: *in* L'Ateneo Parmense, Acta Naturalia, v. 7, fasc. 1, 24 p.

————, 1971, see Selli, R. (ed.), 1971.

Barrell, J., 1917, Rhythms and the measurement of geologic time: Geol. Soc. America Bull., v. 28, p. 745–904.

Bartenstein, H., 1959, Die Jura/Kreide-Grenze in Europa. Ein Überblick des derzeitigen Forschungsstandes: Eclog. Geol. Helv., v. 52, no. 1, p. 15–18.

————, 1965, Unter-Valanginian or Berriasian: Bulgarian Geol. Soc. Rev., v. 26, pt. 1, p. 51–58.

Barthel, K. W., 1964, Die Verteilung der Cephalopoden in den Neuberger Bankkalken, ihr Vergleich mit der Ammonitenfauna von St. Concors und kurze Bemerkungen zum Zonenbegriff: *in* Colloque du Jurassique, Luxembourg (1962), Volume des Comptes Rendus et Mémoires publié par l'Institut grand-ducal, Section des sciences naturelles, physiques et mathématiques, p. 513–517.

———, 1971, Stratigraphic problems; reference sections, the Tithonian, and the Jurassic/ Cretaceous boundary: Neues Jahrb. Geol. Palaeont., Monatsh., no. 9, p. 513–516.

Beede, J. W., and A. F. Rogers, 1908, Coal Measures faunal studies: Faunal divisions of the Kansas Coal Measures: Univ. Geol. Survey of Kansas, v. 9, p. 318–359.

Bell, W. C., 1950, Stratigraphy: a factor in paleontologic taxonomy: Jour. Paleontology, v. 24, p. 492–496.

———, 1959, Uniformitarianism—or uniformity: Am. Assoc. Petroleum Geol. Bull., v. 43, no. 12, p. 2862–2865.

———, 1959, Telling time by tape and type sections: Mimeographed from Program of Am. Assoc. Petroleum Geol. Regional meeting, Lubbock, Texas, October 8–10, 1959.

———, 1960, Review of "Stratigraphic principles and practice" by J. Marvin Weller: Jour. Geol., v. 68, no. 6, p. 684–686.

———, R. R. Berg, and C. A. Nelson, 1956, Croixan type area—upper Mississippi Valley: 20th Int. Geol. Cong. (Mexico), Proc., v. 2, pt. 2, El sistema Cámbrico, su paleografía y el problema de su base, p. 415–446.

———, et al., 1961, Note 25—Geochronologic and chronostratigraphic units: (Prepared by a subcommittee of the American Commission on Stratigraphic Nomenclature): Am. Assoc. Petroleum Geol. Bull., v. 45, no. 5, p. 666–670.

*Belyaevskii, N. A., et al., 1960, Itogi Vsekitaiskogo stratigrafichekogo soveshchaniya (Proceedings of the All-Chinese Stratigraphic Conference, Peking, November 13–14, 1959): Sov. Geol., no. 2.

Benda, L., G. Lüttig, and H. Schneekloth, 1966, Aktuelle Fragen der Biostratigraphie im nordeuropäischen Pleistozän: Eiszeitalter und Gegenwart, v. 17, p. 218–223.

Berger, M. G., 1969, K voprosu o fatsiyakh—k 100 letiyu ucheniyah o fatsiyakh v Rossii (The facies problem—100 years study in Russia): Akad. Nauk SSSR Isv. ser. geol., no. 11, p. 87–94. (English transl., Int. Geol. Rev., v. 12, no. 11, p. 1265–1270, 1970.)

Berggren, W. A., 1962, Stratigraphic and taxonomic-phylogenetic studies of Upper Cretaceous and Paleocene planktonic foraminifera: Stockholm Contributions in Geology, v. 9, no. 2, p. 107–129.

———, 1963, Review and discussion of "Fundamentals of mid-Tertiary stratigraphical correlation" by Eames, Banner, Blow, and Clarke: Micropaleontology, v. 9, no. 4, p. 467–473.

———, 1971, Multiple phylogenetic zonations of the Cenozoic based on planktonic foraminifera: 2nd Planktonic Conf. (Rome, 1970) Proc., (ed. by A. Farinacci), no. 2, v. 1, p. 41–56, Edizioni Tecnoscienza.

———, 1971, Tertiary boundaries and correlations: *in* The Micro-paleontology of the Oceans (ed. by Funnell and Riedell), Cambridge Univ. Press, p. 693–809. (History of Tertiary stratigraphic divisions; lithostratigraphical, biostratigraphical and chronostratigraphical terminology and usage; stratigraphic principles), p. 693–702.

———, 1971, Neogene chronostratigraphy, planktonic foraminiferal zonation and the radiometric time scale: Földtani Közlony, Hungarian Geol. Soc. Bull. 101, p. 162–169. (Colloquium on the Neogene, Budapest, 1969, September 4–9).

———, 1972, A Cenozoic time-scale—some implications for regional geology and paleobiogeography: Lethaia, v. 5, p. 195–215.

———, 1973, The Pliocene time scale: calibration of planktonic foraminiferal and calcareous nannoplankton zones: Nature, v. 243, no. 5407, p. 391–397.

———, 1973, Biostratigraphy and biochronology of the Late Miocene (Tortonian and Messinian) of the Mediterranean: *in* Messinian events in the Mediterranean, Koninklijke Neder. Akad. Wetensch., Amsterdam, p. 10–20.

——— et al., 1967, Late Pliocene-Pleistocene stratigraphy in deep sea cores from the south-central North Atlantic: Nature, v. 216, October 21, 1967, p. 253–254.

Berry, W. B. N., 1962, Chorology, chronology, and correlation: *in* Geol. Soc. America Special Paper 68 (Abstracts for 1961), p. 134–135.

———, 1966, Zones and zones—with exemplifications from the Ordovician: Am. Assoc. Petroleum Geol. Bull., v. 50, no. 7, p. 1487–1500.

———, 1968, Growth of a prehistoric time scale: Freeman, 158 pp.

———, 1974, Erben's "Inventory in Štratigraphy"—a model from the California Tertiary foraminifer succession: Newsl. Strat., v. 3, no. 2, p. 65–72.

——— and **A. J. Boucot,** 1970, Correlation of the North American Silurian rocks: Geol. Soc. America Spec. Paper no. 102, 289 pp.

* **Bertenshteĭn, Kh.,** 1966, K primeneniiu stratigraficheskoĭ i khronologicheskoĭ terminologii v mikropaleontologii (Applications of stratigraphic and chronologic terminology in micro-paleontology): Voprosy mikropaleontologii, vyp. 10, Izd. "Nauka."

Bertolino, V., et al., 1968, Proposal for a biostratigraphy of the Neogene in Italy based on planktonic foraminifera: Giorn. Geol., Museo Geol. Bologna Ann., ser. 2, v. 35, fasc. II, p. 23–30.

Beurlen, K., 1963, O têrmo formação na terminologia estratigráfica, ilustrado pelas Formações Maruim e Gramame (Cretáceo do Nordeste do Brasil): Acad. Brasileira Ciencias Anais, v. 35, p. 327–338.

Biquand, D., 1972, Application du paléomagnétisme à la resolution de problèmes stratigraphiques: difficultés et limites actuelles de la méthode: *in* Colloque sur les méthodes et tendances de la stratigraphie (Orsay, 1970), BRGM France, Mém. 77, pt. 2, p. 861–876.

Blackwelder, E., 1924, Suggestions for the improvement of our geologic terminology (abstr.): Geol. Soc. America Bull., v. 35, no. 1, p. 103, Pan-American Geol., v. 41, no. 2, p. 151.

Blanford, W. T., 1884, On the classification of sedimentary strata: Geol. Mag., 3rd ser., v. 1, p. 318–321.

———, 1889, The anniversary address of the president: Geol. Soc. London Quart. Jour. v. 45, Proc., p. 37–77.

Bliss, N. W., 1968, The need for a revised stratigraphic nomenclature in the Precambrian of Rhodesia: Annexure to v. 71, Symposium on the Rhodesian Basement Complex—Geol. Soc. South Africa, Rhodesian Branch, p. 205–213.

Blondeau, A., and C. Pomerol, 1968, Qu'est-ce que L'Auversien?: BRGM France, Mém. 58, (Colloque sur L'Eocène, Paris, May 1968), p. 565–574.

Blow, W. H., 1969, Late middle Eocene to recent planktonic foraminiferal biostratigraphy: *in* 1st Int. Conf. on Planktonic Microfossils (Geneva, 1967), Proc., v. 1 (ed. by Bronnimann and Renz), E.J. Brill, Leiden, p. 199–422.

————, 1970, Validity of biostratigraphic correlations based on the *Globigerinacea*: Micropaleontology, v. 16, no. 3, p. 257–268.

————, 1971, Geostratigraphy and a philosophical basis for the interpretation of the geohistorical record: (Manuscript), 12 p.

————, 1971, Principles, nomenclature and philosophy of biostratigraphy: (Manuscript), 15 p.

————, and **F. T. Banner,** 1965, A review of stratigraphic terminology: London Palaeontological Note no. 502, (Manuscript), 31 p.

*Bodylevsky, V. I., 1964, On the stratigraphic zone (Title translated from Russian): Trudi, VSEGEI, v. 102, p. 25–32.

Bogolepov, K. V., 1970, Nekotoryye voprosy ucheniya o geologicheskikh formatsiyakh (Some problems of the theory of geologic formations): Geol. i geofiz., no. 1, p. 39–49. (English trans., Int. Geol. Rev., v. 12, no. 12, p. 1502–1508.)

Bogsch, L., 1962, Einige prinzipielle und praktische Fragen der erdgeschichtlichen Grenzen auf Grund egerer Fauna: Annal. Univ. Sci. Budapest, Sect. Geol., v. 5 (1961), p. 11–23.

Bokman, J., 1956, Terminology for stratification in sedimentary rocks: Geol. Soc. America Bull., v. 67, no. 1, p. 125–126.

Bolli, H. M., 1966, Zonation of Cretaceous to Pliocene marine sediments based on planktonic foraminifera: Bol. Informativo, Assoc. Venezolana de Geol. Min., Pet., v. 9, no. 1, p. 3–32.

————, 1969, Report of Working Group for a biostratigraphic zonation of the Cretaceous and Cenozoic as a basis for correlation in marine geology: IUGS Geol. Newsl., v. 1969, no. 3, p. 199–207.

————, 1970, Information for a review of the activities of the Working Group for Correlation of Cretaceous and Cenozoic Marine Deposits to be given at 4th general meeting of Commission for Marine Geology at 15th General Assembly of Int. Assoc. Phys. Sci. of the Oceans, September 1970, Tokyo: Multilith copy, 5 p.

Bombita, G., and **V. Moisescu,** 1968, Données actuelles sur le nummultique de Transylvanié: Colloque sur l'Eocene, BRGM France, Mém. 58, p. 693–729.

Bomboe, P., and **A. Maries,** 1972, An algorithm for stratigraphical groupings (abstr.): 24th Int. Geol. Cong. (Montreal), Abstracts, p. 519–520.

Boni, A., 1951, Il tempo nelle scienze geologiche: Istit. Geol. Univ. Pavia Atti., v. 4 (1950), p. 3–25.

*Borovikov, L. I., and **T. N. Spizharsky,** 1965, Principles of subdivision and correlation of Pre-Cambrian deposits (Title translated from Russian): Geol. i geofiz. no. 1, p. 21–29.

Borrello, A. V., 1965, Sistemática estructural sedimentaria en los procesos de la orogenesis: Com. Invest. Cient. An. (Provincia Buenos Aires), v. 6, p. 65–93.

———— and **A. J. Cuerda,** 1963, Sobre el código de nomenclatura estratigráfica y su significación: Com. Invest. Cient. An. (Provincia Buenos Aires), v. 4, p. 515–521.

Boškov-Štajner, Z., 1968, Preliminarni stratigrafski kodeks: Nafta, Zagr., v. 19, no. 12, p. 529–534. (English transl. by **Darinka Zigic Briggs,** Ann Arbor Michigan, November 1970, manuscript, 17 p., made for U.S. Geol. Survey.)

————, 1968, Stratigraphic units of the southern part of Pannonian Basin in the territory of Yugoslavia: Bull. Scientifique, section A, v. 13, nos. 3–4, p. 73–74.

——, 1969, Geologija Mramor-Brda (1): Nafta, v. 20, no. 6, p. 279–287.

——, 1974, Stratigrafski principi, nomenklatura, terminologija i stratotipovi prema uputama internacionalne potkomisije za stratigrafsku klasifikaciju (ISSC) (Stratigraphic principles, nomenclature, terminology and stratotypes according to International Subcommission on Stratigraphic Classification (ISSC): Nafta, Zagreb, v. 25, no. 6, p. 298–300.

—— and **T. Tomasović,** 1967, Stratigrafska pripadnost naftnih i plinskih kolektora u SR Hrvatskoj (The stratigraphic age of oil and gas reservoir rocks in the S.R. Croatia): VI Geoloski kongres Jugoslavije, Ohrid.

—— **et al.,** 1967, Stratigraphic units of southern part of the Pannon Basin in the territory of the Soc. Fed. Rep. of Yugoslavia: Com. Med. Neogene Strat., Proc. 4th Session. 1969, Giorn. Geol., ser. 2, v. 35, fasc. 4, p. 287–296.

—— and **D. Marinovic,** 1971, Stratigraphy of oil and gas fields in the territory of Yugoslavia: Nafta, v. 22, no. 6, p. 524–532.

Boucek, B., R. Horny, and **I. Chlupáč,** 1962, Diskussion zur Silur/Debon-Grenze: *in* Symposium Silur/Devon-Grenze (1960), Stuttgart, p. 304–305.

Boucot, A. J., 1970, Practical taxonomy, zoogeography, paleoecology, paleogeography and stratigraphy for Silurian and Devonian brachiopods: *in* Correlation by fossils, North American Paleont. Conv., 1969, Proc., pt. F, p. 566–611.

Boué, A., 1830–1831, Classifications: Geol. Soc. France Bull., v. 1, p. 107–113.

Bourbeau, G. A., 1958, Soils in stratigraphic nomenclature: Am. Assoc. Petroleum Geol. Bull., v. 42, no. 8, p. 1987–1992.

Bouroz, A., 1972, Synthèse de la section marqueurs volcaniques. Utilisation des marqueurs d'origine volcanique en stratigraphie: *in* Colloque sur les méthodes et tendances de la stratigraphie, Orsay (1970), BRGM France, Mém. 77, pt. 1, p. 461–465.

Boussac, J., 1910, Du role de l'hypothèse en paleontologie stratigraphique: Revue Scientifique, Paris, 48th ann., p. 5–9.

——, 1912, Études stratigraphiques sur le Nummulitique Alpin: Mém. Carte Geol. France, p. xi–xvii.

Bradshaw, J. D., 1968, New Zealand Permian stages: New Zealand Jour. Geol. Geophys., v. 11, no. 1, p. 265–267.

Bramlette, M. N., 1948, Discussion of nature and classes of stratigraphic units: Am. Assoc. Petroleum Geol. Bull., v. 32, no. 3, p. 381.

——, 1965, Massive extinctions in biota at the end of Mesozoic time: Science, v. 148, June 25, 1965, p. 1696–1699.

—— and **F. R. Sullivan,** 1961, Coccolithophorids and related nannoplankton of the early Tertiary in California: Micropaleontology, v. 7, no. 2, p. 129–188.

—— and **J. A. Wilcoxon,** 1967, Middle Tertiary calcareous nannoplankton of the Cipero section, Trinidad, W.I.: Tulane Studies in Geology, v. 5, no. 3, p. 93–131.

—— and **W. R. Riedel,** 1971, Observations on the biostratigraphy of pelagic sediments: *in* The micropaleontology of oceans, Cambridge Univ. Press, p. 665–668.

Branson, C. C., 1956, Cyclic formations or mappable units: Oklahoma Geology Notes, v. 16, p. 122–126.

——, 1961, Code of stratigraphic nomenclature: Oklahoma Geology Notes, v. 21, no. 12, p. 317–322.

Breddin, H., 1938, Bemerkungen zur Frage der Richtprofile: Deutsche Geol. Gesell Zeit., v. 90, p. 231–232, Berlin.

————, 1962, Die naturwissenschaftliche Methodik in der Geologie: Geol. Mitt., v. 3, p. 23–32.

Bretz, J., et al., 1959, Discussion of Report-6—Application of stratigraphic classification and nomenclature to the Quaternary: Am. Assoc. Petroleum Geol. Bull., v. 43, no. 3, p. 674–675.

Brewer, R., 1972, Use of macro- and micromorphological data in soil stratigraphy to elucidate surficial geology and soil genesis: Geol. Soc. Australia Jour., v. 19, pt. 3, p. 331–344.

————, **K. A. W. Crook,** and **J. G. Speight,** 1970, Proposal for soil-stratigraphic units in the Australian stratigraphic code: Geol. Soc. Australia Jour., v. 17, no. 1, p. 103–111.

Brinkmann, R., 1928, Statistisch-phylogenetische Untersuchungen an Ammoniten: 5th Int. Kong. Vererbungswissensch. (5th Int. Cong. of Genetics), Berlin, 1927, Verh.: suppl.-bd. no. 1, Zeit. Indukt. Abstamm. u. Vererbungslehre, p. 496–513.

British National Committee for Geology, 1968, International Geological Correlation Program, United Kingdom contribution: Royal Soc. London, 43 p. (Contains revised edition of 1967 Report of the Stratigraphic Code Sub-committee).

Broeck, E. van den, 1883, Note sur un nouveau mode de classification et de notation graphique des dépôts géologiques: Musée Royal d'histoire naturelle de Belgique Bull., v. 2, p. 341–369.

Broecker, W. S., and **J. I. Kulp,** 1956, The radiocarbon method of age determination: Am. Antiquity, v. 22, no. 1, p. 1–11.

Brongniart, A., 1829, Tableau des terrains qui composent l'écorce du globe, ou Essai sur la structure de la partie connue de la terre: Paris, 433 p.

Bronnimann, P., and **J. Resig,** 1971, A Neogene Globigerinacean biochronologic time-scale of the southwestern Pacific: Initial reports of the Deep Sea Drilling Project, v. 7, pt. 2, p. 1235–1469, Washington.

Brooks, J. E., and **D. L. Clark,** 1961, Thermoluminescence as a correlation tool in the Austin Chalk in north central Texas: Graduate Research Center, Southern Methodist Univ. Jour., v. 29, no. 3, p. 198–204.

Brown, L. F., 1959, Problems of stratigraphic nomenclature and classification, upper Pennsylvanian, north central Texas: Am. Assoc. Petroleum Geol. Bull., v. 43, no. 12, p. 2866–2871.

Brunn, J. H., 1972, Reflexions sur les objectifs de la stratigraphie et les moyens qu'elle met en oeuvre: *in* Colloque sur les méthodes et tendances de la stratigraphie (Orsay, 1970), BRGM France, Mém. 77, pt. 2, p. 1001–1005.

Brunnschweiler, R. O., 1949, Additional notes on stratigraphic nomenclature concerning the term "zone": Mimeographed sheet received from Australian Geol. Survey.

*****Bublichenko, N. L.,** 1962, On methods of stratigraphic investigations in Rudny Altai (Title translated from Russian): Trudi Altaiskogo gornometal. Inst., v. 12, p. 3–21.

Bubnoff, S. von, 1963, Fundamentals of geology (English translation of Grundprobleme der Geologie, 1954): Oliver and Boyd, Edinburgh, 287 p.

Buch, L. von, 1810, Etwas über locale und allgemeine Gebirgsformationen: Gesellschaft naturforschender Freunde zu Berlin, Jahrg. 4, p. 69–74.

Buckman, S. S., 1893, The Bajocian of the Sherborne district: its relation to subjacent and superjacent strata: Geol. Soc. London Quart. Jour., v. 49, p. 479–522.

————, 1898, On the grouping of some divisions of so-called Jurassic time: Geol. Soc. London Quart. Jour., v. 54, p. 442–462.

————, 1902, The term "Hemera": Geol. Mag., new ser., v. 9, p. 554–557.

————, 1903, The term "Hemera": Geol. Mag., new ser., v. 10, p. 95–96.

Bukry, D., 1971, Cocolith stratigraphy, Leg 7, Deep Sea Drilling Project: Initial reports of the Deep Sea Drilling Project, v. 7, pt. 2, p. 1513–1528, Washington.

————, 1973, Cocolith stratigraphy, eastern equatorial Pacific, Leg 16, Deep Sea Drilling Project: Initial reports of the Deep Sea Drilling Project, v. 16, p. 653–711, Washington.

———— **et al.,** 1971, Planktonic microfossil biostratigraphy of the northwestern Pacific Ocean: Initial reports of the Deep Sea Drilling Project, v. 6, p. 1253–1300, Washington. (See p. 1294 for discussion of the relation to acoustostratigraphy.)

Burek, P. J., 1968, Korrelation reversmagnetisierter Gesteinsfolgen als stratigraphisches Hilfsmittel-aufgezeigt am Beispiel einer paleomagnetischen Studie im Buntsandstein SW-Deutschlands: 23rd Int. Geol. Cong. (Prague), v. 5, p. 23–36.

Burollet, P. F., 1956, Contribution à l'étude stratigraphique de la Tunisie Centrale: Annales des Mines et de la Géologie, no. 18, 345 p.

————, 1959, Remarques sur la nomenclature stratigraphique: Sciences de la Terre, v. 5 (1957), nos. 2–3, p. 117–136.

Busson, G., 1972, Nomenclature et classification stratigraphiques: confrontation des problèmes actuels aux données de l'étude de Mésozoïque saharien: p. 45–82, Chapter 4, Part 1 of Principes, méthodes et résultats d'une étude stratigraphique du Mésozoïque saharien, Mus. Natl. Hist. Nat. (Paris) Mem., Ser. C., v. 26, 441 p.

Cahen, L., 1958, Quelques considerations sur les relations entre Precambrien et Cambrien et le problème des séries intermédiares: Colloques Int. du Centre Nat. Rech. Sci., Paris, 1957, p. 133–138.

Calkins, F. C., 1941, "Band", "layer", and some kindred terms: Econ. Geol., v. 36, no. 3, p. 345–349.

Callomon, J. H., 1965, Notes on Jurassic stratigraphical nomenclature 1. Principles of stratigraphic nomenclature: Carpatho-Balkan Geological Assoc., VII Cong., Sofia, September 1965, Reports, part II, v. 1, p. 81–89.

———— and **D. T. Donovan,** 1966, Stratigraphic classification and terminology: Correspondence, Geol. Mag., v. 103, no. 1, p. 97–99.

———— and **D. T. Donovan,** 1971, A code of Mesozoic stratigraphical nomenclature: BRGM France, Mém. 75 (Colloque du Jurassique, Luxembourg, 1967), p. 75–81.

Carozzi, A. V., 1951, La notion de synchronisme en géologie: Rev. gen. sci. pures et appl., v. 58, nos. 7–8, p. 230–236.

Carter, R. M., 1970, A proposal for the subdivision of Tertiary time in New Zealand: New Zealand Jour. Geol. Geophys., v. 13, no. 2, p. 350–363.

————, 1974, A New Zealand case-study of the need for local time-scales: Lethaia, v. 7, no. 3, p. 181–202.

———— **et al.,** 1973, Suggestions towards a high-level nomenclature for New Zealand rocks: Royal Soc. New Zealand Jour., v. 4, no. 1, p. 5–18.

Castellarin, A., 1969, Report on the discussions during the session on stratigraphy and paleogeography: in Com. Med. Neogene Strat., Proc. 4th session, Bologna (1967), pt. 4, Giorn. Geol. ser. 2a, v. 35, no. 4, p. 231–233.

Caster, K. E., 1934, The stratigraphy and paleontology of northwestern Pennsylvania, Part I, Stratigraphy: Bull. Am. Pal., v. 21, 185 p.

Catalano, R., and V. Liguori, 1971, Facies a Calpionelle della Sicilia occidentale: 2nd Planktonic Conf. (Rome, 1970) Proc., (ed. A. Farinacci) Edizioni Tecnoscienza, p. 167–209.

―――― and R. Sprovieri, 1971, Biostratigrafia di alcune serie Saheliane (Messiniano Inferiore) in Sicilia: 2nd Planktonic Conf. (Rome, 1970) Proc., (ed. A. Farinacci) Edizioni Tecnoscienza, p. 211–249.

Cati, F., et al., 1968, Biostratigrafia del Neogene mediterraneo basata sui foraminiferi planctonici: Soc. Geol. Ital. Boll., 87, p. 491–503.

Chadwick, G. H., 1930, Subdivision of geologic time: Geol. Soc. America Bull., v. 41, p. 47–48.

Challinor, J., 1967, A dictionary of geology: 3rd ed., Univ. of Wales, Cardiff, 298 p.

Chamberlin, R. T., 1935, Certain aspects of geologic classification and correlation: Science, v. 81, p. 183–190 and 216–218.

Chamberlin, T. C., 1898, The ulterior basis of time divisions and the classification of geologic history: Jour. Geology, v. 6, p. 449–462.

―――― and R. D. Salisbury, 1906, Geology, v. 2 (Earth History), Holt, New York, 692 p.

Chang, K. H., 1968, A review of stratigraphic classification (with emphasis on the classification of Korean stratigraphy): Volcano, v. 10, no. 1, p. 5–13.

――――, 1973, Toward an international guide to stratigraphic classification, terminology, and usage: Geol. Soc. Korea Jour., v. 9, no. 2, p. 123–125.

――――, 1974, Origin of multiple stratigraphic classification and an unpublished 1932 manuscript of H.D. Hedberg: Geol. Soc. America Bull., v. 85, no. 8, p. 1301–1304.

Cheetham, A. H., and P. B. Deboe, 1963, A numerical index for biostratigraphic zonation in the Mid-Tertiary of the eastern Gulf: Gulf Coast Assoc. Geol. Soc., Trans., v. 13, p. 139–147.

Chiji, M., 1961, Neogene biostratigraphy of the Toyama sedimentary basin, Japan Sea coast (in Japanese with English summary): Osaka Mus. Nat. Hist. Bull., no. 14, 88 p.

Childs, T. S., et al., 1941, Letter to Dr. C. W. Tomlinson on stratigraphic nomenclature: Am. Assoc. Petroleum Geol. Bull., v. 25, no. 12, p. 2195–2211.

China, 1960, Chinese code of stratigraphic nomenclature (in Chinese): Geol. Soc. China, Proc., no. 3, p. 2–5. (Fide Biq Chingchang).

* China (Peoples Republic of China Stratigraphic Conference), 1965, Project of a stratigraphic code: Pekin, 54 p.

China, The Geological Society of, 1970, Note on establishment of principles for standardization of stratigraphic nomenclature in Taiwan, China: in Stratigraphic correlation between sedimentary basins of the ECAFE region, U.N. Econ. Commission Asia Far East, v. 2 of Mineral Resources Development Series, no. 36, p. 53.

Chlupáč, I., 1957, Zásady stratigrafické terminologie v SSSR: Věstnik Ústř. Úst. Geol., v. 32, no. 5, p. 301–308, Praha; (French transl. no. 1623 by E. Jayet, SIG, Paris: Principes de la terminologie stratigraphique en SSSR, 10 p.)

――――, 1963, Vydání mezinárodních zásad stratigrafické klasifikace a terminologie (The edition of principles of stratigraphical classification and terminology): Věstnik Ústř. Úst. Geol., v. 38, p. 61–63, Praha.

――――, 1966, Mezinárodni diskuse o vymezaní geologických útvaru (International discussion of the delimitation of geological Systems): Věstník Ústř. Úst. Geol., v. 41, p. 1–7.

————, 1968, Prípava návrhu mezinárodní litostratigrafické klasfikace (The proposal of international lithostratigraphic classification): Věstník Ústř. Úst. Geol., v. 43, p. 1–7.

————, 1969, Soucasný stav základního stratigrafického deleni (Present knowledge of basic stratigraphic subdivisions): Casopis Mineral Geol., v. 14, no. 3–4, p. 249–257.

————, 1970, Chronostratigraphy and neostratotypes. Comments by Czechoslovak Stratigraphic Commission: Am. Assoc. Petroleum Geol. Bull., v. 54, no. 7, p. 1317.

————, 1973, Současné směry ve stratigrafické geologii (Present concepts in stratigraphical geology): Věstník Ústř. Úst. Geol., v. 48, no. 2, p. 65–71, Praha.

————, **H. Jaeger,** and **J. Zikmundova,** 1972, The Silurian-Devonian boundary in the Barrandian: Bull. Canadian Petroleum Geol., v. 20, no. 1, p. 104–174.

Cicha, I., et al., 1964, Project provisoire pour une subdivision chronostratigraphique du Tertiaire: BRGM France, Mém. 28 (Colloque sur le Paléogène, Bordeaux, 1962), p. 925–929.

————, **J. Seneš, J. Tejkal, et al.,** 1967, Chronostratigraphie und Neostratotypen, Miozän der Zentralen Paratethys, Bd.I, M₃ (Karpatien) Die Karpatische Serie und ihr Stratotypus, Vydavateľstvo Slovenskej Akad. vied Bratislava, 312 p.

————, **J. Seneš, J. Tejkal, et al.,** 1969, Proposition pour la création de néostratotypes et l'établissement d'une échelle chronostratigraphique dite ouverte: *in* Com. Med. Neogene Strat. Proc., 4th session (Bologna, 1967): pt. 4, Giorn. Geol., ser. 2a, v. 35, no. 4, p. 297–311.

————, **J. Seneš,** and **J. J. Tejkal,** 1969, Summary of section on chronostratigraphy and neostratotypes: Am. Assoc. Petroleum Geol. Bull., v. 53, no. 10, p. 2204–2206.

————, and **J. Seneš,** 1971, Probleme der Beziehung zwischen Bio- und Chronostratigraphie des jüngeren Tertiäre: Geol. zb.-Geol. Carpathica (Slov. Akad. Vied) v. 22, no. 2, p. 209–228.

Cita, M. B., 1971, Definition and use of the Bormidian Stage. A critical review: Newsl. Strati., v. 1, no. 4, p. 29–43.

————, 1971, See Selli, R. (ed.), 1971.

————, 1973, Inventory of biostratigraphical findings and problems: Initial reports of the Deep Sea Drilling Project, v. 13, p. 1045–1065, Washington.

————, 1973, Pliocene biostratigraphy and chronostratigraphy: Initial reports of the Deep Sea Drilling Project, v. 13, p. 1343–1379, Washington.

Cline, M. G., 1949, Basic principles of soil classification: Soil Sci., v. 67, p. 81–91.

Cloud, P., 1971, Precambrian of North America (The Third Penrose Conference): Geotimes, v. 16, no. 3, p. 13–18.

————, 1973, Possible stratotype sequences for the basal Paleozoic in North America: Am. Jour. Sci., v. 273, no. 3, p. 193–206.

Cohee, G. V., 1960, Series subdivisions of Permian System: Am. Assoc. Petroleum Geol. Bull., v. 44, no. 9, p. 1578–1579.

————, 1962, Stratigraphic nomenclature in reports of the U.S. Geological Survey: U.S. Geol. Survey, Washington, 35 p.

———— (chairman, Committee on Standard Stratigraphic Coding), 1967, Standard stratigraphic code adopted by AAPG: Am. Assoc. Petroleum Geol. Bull., v. 51, no. 10, p. 2146–2150.

————, 1968, Holocene replaces Recent in nomenclature usage of the U.S. Geological Survey: Am. Assoc. Petroleum Geol. Bull., v. 52, no. 5, p. 582.

―――, 1970, Stratigraphic nomenclature; principles and procedures: *in* Geol. Seminar on the North Slope of Alaska, Am. Assoc. Petroleum Geol., Proc., Pacific Section, p. H1–H3, Los Angeles.

―――, 1972, Reports published by International Subcommission on Stratigraphic Classification—a review: Am. Assoc. Petroleum Geol. Bull., v. 56, no. 6, p. 1135–1136.

――― (ed.), 1974, Stratigraphic nomenclature in reports of the U.S. Geological Survey: U.S. Dept. Interior, Geol. Surv., 0–585–465/1, 45 p.

――― and **J.B. Patton,** 1963, Discussion of the stratigraphic code: capitalization: Am. Assoc. Petroleum Geol. Bull., v. 47, no. 5, 852–853.

―――, **R. K. DeFord,** and **H. B. Willman,** 1969, Note 36 (of Am. Comm. Strat. Nomen.)—Amendment of Article 5, Remarks (a) and (e) of the Code of Stratigraphic Nomenclature for treatment of geologic names in a gradational or interfingering relationship of rock-stratigraphic units: Am. Assoc. Petroleum Geol. Bull., v. 53, no. 9, p. 2005–2006.

Colalongo, M. L. et al., 1972, Biostratigrafia e cronostratigrafia del Pliocene: Soc. Geol. Ital. Boll., v. 91, p. 489–509.

Collins, B. W., 1945, Review of "Stratigraphical classification and nomenclature" by F. R. S. Henson: Am. Assoc. Petroleum Geol. Bull., v. 29, no. 8, p. 1208–1211.

Colloque sur l'Eocéne (1968), 1969, Propositions: BRGM France, Mém. 69, p. 459–470. (In French and English).

Colloque sur le Crétacé inférieur (Lyon, September 1963), 1965, Les étages du crétacé inférieur étudiés à partir des stratotypes: BRGM France, Mém. 34, 840 p. (Many authors).

Colloque sur les méthodes et tendances de la stratigraphie (Orsay, 1970), 1972, BRGM France, Mém. 77, 2 parts, 1011 p. (Contains numerous articles related to the subject of this bibliography of which only a few are listed here by authors.)

Commission on Geochronology (IUGS), 1968, Vergleichende Tabelle der letzten veröffentlichten geochronologischen Zeitskalen der phanerozoischen Epochen: Zeit. Angew. Geol., v. 14, no. 8, p. 440–442.

Conkin, J. E., and **B. M. Conkin,** 1973, The paracontinuity and the determination of the Devonian-Mississippian boundary in the type Lower Mississippian area of North America: Univ. of Louisville Studies in Paleontology and Stratigraphy no. 1, 36 p.

Conybeare, W. D., and **W. Phillips,** 1822, Outlines of the geology of England and Wales: London, 470 p. (See introduction, p. 1–61.)

Cook, H. E., 1972, North American stratigraphic principles as applied to deep-sea sediments (abstr.): Am. Assoc. Petroleum Geol. Bull., v. 56, no. 3, p. 609–610.

―――, 1972, Stratigraphy and sedimentation: Initial reports of the Deep Sea Drilling Project, v. 9, p. 933–943, Washington.

―――, 1975, North American stratigraphic principles as applied to deep-sea sediments: Am. Assoc. Petroleum Geol. Bull., v. 59 no. 5, p. 817–837.

Cooper, B. N., 1960, Systemic boundaries in the Appalachians: Mineral Industries Jour., Virginia Polytechnic Inst., v. 7, no. 4, p 5–8.

Cooper, W. S., 1958, Terminology of post-Valders time: Geol. Soc. America Bull., v. 69, no. 7, p. 941-945.

Cotton, C. A., 1950, Discordant time scales: Science, v. 111, p. 14–15.

Córdoba, D. A., and **Z. de Cserna,** 1970, Código de nomenclatura estratigráfica por la Comisión Americana de Nomenclatura Estratigráfica: Spanish translation of 2nd edition (1970) of American Code of Stratigraphic Nomenclature, Mexico, D.F., 28 p.

Crescenti, U., 1966, Sulla biostratigrafia del Miocene affiorante al confine Marchigiano-Abruzzese: Geologica Romana, v. 5, p. 1–54.

Crook, K. A. W., 1962, A note on stratigraphical nomenclature—biostratigraphic zones and time-rock stages: Royal Soc. New South Wales, v. 96, p. 15–16.

———, 1966, Principles of Precambrian time-stratigraphy: Geol. Soc. Australia Jour., v. 13, pt. 1, p. 195–202.

Cross, W., 1902, Geologic formations versus lithologic individuals: Jour. Geology, v. 10, no. 2, p. 223–244.

Cserna, Z. de, 1972, Essay review of 'Stratigraphie und Stratotypus' by O.H. Schindewolf: Am. Jour. Sci., v. 272, no. 2, p. 189–194.

Cumings, E. R., 1932, Reefs or bioherms: Geol. Soc. America Bull., v. 43, p. 331–352.

Cumming, A. D., J. G. C. M. Fuller, and **J. W. Porter,** 1959, Separation of strata: Paleozoic limestones of the Williston Basin: Am. Jour. Sci., v. 257, p. 722–733.

Curry, D., 1967, Problems of correlation in the Anglo-Paris-Belgian Basin: Geol. Assoc. London, Proc., v. 77, p. 437–467, (1966).

Czechoslovakia, 1960, Ceskoslovenská stratigrafická terminologie (Czechoslovak stratigraphic terminology): Věstník Ústř. Úst. Geol., v. 35, p. 95–110, Prague.

Czechoslovakian National Committee on Stratigraphy, 1960, Prager Arbeitstagung über die Stratigraphie des Silurs und des Devons (1958) — Beschluss: Edit. Ústř. Úst. Geol., Praha, p. 509–513.

Dagley, P., et al., 1967, Geomagnetic polarity zones for Icelandic lavas: Nature, v. 216, no. 5110, p. 25–29.

Dalbiez, M. F., 1959, Rapport V, Correlations et Resolutions 81st Congrès des Soc. Savantes, Colloque sur le Crétace supérieur Français, p. 857–867.

Dana, J. D., 1856, On American geological history: Am. Jour. Sci., 2nd ser., v. 22, no. 66, p. 305–334.

———, 1863, Manual of Geology: Philadelphia, 798 p. (see p. 90–134); 2nd ed., New York, 1875, 828 p. (see p. 79–146); 3rd ed., New York, 1880, 911 p.; 4th ed., New York, 1895, 1088 p.

Daniels, S. G. H., and **S. J. Freeth** (editors), 1970, Stratigraphy: an interdisciplinary symposium: Ibadan Univ., Inst. African Studies, Occasional Publications, no. 19, 56 p.

Davitashvili, L. C., 1943, Sur les méthodes d'une subdivision chronologique des dépôts du Tertiaire de la province pétrolifère du Caucase: Tr. Gruz. Ind. in-ta im. S.M. Kirova, no. 1 (15), (Fide Stepanov as translated to French by Mme. Jayet SIG, Paris.)

———, 1948, Bases théoriques de la synchronisation des dépôts du Tertiaire supérieur du bassin de la Caspienne - Mer Noire: Mat VSEGEI, Paléont. i strat., sb. 5. (Fide Stepanov as translated to French by Mme. Jayet SIG, Paris.)

* **Debrowsky, Z.,** 1963, Uwagi a nomenklaturze stratigrafii malma: Przegl. Geol., v. 11, no. 5, p. 241–243.

DeFord, R. K., 1957, Discussion of Report 5 (of Am. Comm. Strat. Nomen.) — Nature, usage, and nomenclature of biostratigraphic units: Am. Assoc. Petroleum Geol. Bull., v. 41, no. 8, p. 1887.

————, **J. A. Wilson,** and **F. M. Swain,** 1967, Note 35 (of Am. Comm. Strat. Nomen.) — Application to American Commission on Stratigraphic Nomenclature for an amendment of Article 3 and Article 13, Remarks (c) and (e) of the Code of Stratigraphic Nomenclature to disallow recognition of new stratigraphic names that appear only in abstracts, guidebooks, microfilms, newspapers, or in commercial trade journals: Am. Assoc. Petroleum Geol. Bull., v. 51, no. 9, p. 1868–1869.

Demarcq, G., 1967, Stratotypes et biostratigraphie du Neogène, essai de méthodologie: Trav. Lab. Géol. Fac. Sci., Univ. Lyons, n.s., no. 14, p. 29–32.

————, 1969, De l'inutilité d'un nouvel étage entre Miocène ét Pliocène: *in* Com. Med. Néogene Strat. 4th Session: Proc. pt. 4, Giorn. Geol. (Bologna), ser. 2a, v. 35, no. 4 (1967), p. 3–6.

Deraniyagala, P. E. P., 1958, The Pleistocene of Ceylon: Colombo National Museum, 164 p.

* **Derviz, T. L.,** 1968, Neobkhodima geokhronologicheskaia stratigrafiia (Geochronologic stratigraphy): Geol. i Geofiz., no. 5.

Deshayes, G. P., 1830–31, Tableau comparatif des espèces de coquilles vivantes, etc.: Géol. Soc. France Bull., v. 1, no. 8, p. 185–187.

Desnoyers, J., 1829, Observations sur un ensemble de dépôts marins plus récents que les terrains tertiaires du bassin de la Seine, et constituant une formation géologique distincte, précédées d'un aperçu de la non-simultanéité des bassins tertiaires: Anal. Sci. Nat., v. 16, p. 171–214, 402–491.

Dewalque, C., et al., 1888, Rapports de la Commission pour l'uniformité de la nomenclature: Int. Geol. Cong. (Berlin), (1885), C.R. 3rd Session, p. 317–399.

Diener, C., 1909, Summary: *in* Krafft, A.v., and C. Diener, Lower Triassic Cephalopoda from Spiti, Malla Johar, and Byans: Geol. Survey India (Pal. Ind.) ser. 15, v. 6, Mem. 1, 186 p. (see p. 163–186.)

————, 1918, Die Bedeutung der Zonengliederung für die Frage der Zeitmessung in der Erdgeschichte: Neues Jahrb. für Min., Geol., und Palaeont., Beilage- Bd. 42, p. 65–172.

————, 1925, Grundzüge der Biostratigraphie: Deuticke, Leipsig, 304 p.

Dineley, D. L., 1964, The chronological value of fossils: *in* Geochronology in Canada, Royal Soc. Canada Spec. Pub. no. 8, Univ. Toronto Press, p. 9–19.

Dollo, L., 1909, La paléontologie éthologique: Soc. Bèlge Géol. Bull., v. 23, p. 377–421.

Donovan, D. T., 1966, Stratigraphy, an introduction to principles: Murby, London, 199 p.

Donovan, R. N., R. J. Foster, and **T. S. Westoll,** 1974, A stratigraphical revision of the Old Red Sandstone of North-eastern Caithness: Royal Soc. Edinburgh, Trans., v. 69, no. 8, p. 167–201.

Dorf, E., 1969, Paleobotanical evidence of Mesozoic and Cenozoic climatic changes: Proc. North American Paleont. Convention, September 1969, pt. D, p. 323–346.

Dott, R. H. Jr., and **R. L. Batten,** 1971, The relative geologic time scale and modern stratigraphic principles: *in* Evolution of the Earth, McGraw-Hill, New York, 649 p. (See p. 53–75.)

Dragunov, V. I., 1962, On building of geohistorical scale of evolution of the earth's crust in connection with problem of stratigraphy, taxonomy and nomenclature of the late Pre-Cambrian deposits (Title translated from Russian): *in* Sovescanie po problemam astrogeologii, Leningrad, p. 148–151.

Dreyfuss, M., 1953, La notion d'étage géologique et les variations locales de la subsidence: Soc. Hist. Nat. du Doubs Bull., no. 57, p. 105–110.

————, 1962, Réflexions sur quelques "unités" employées en stratigraphie et en paléontologie: Bull. Trimestriel du Dépt. d'Information Géologique, 14th year, no. 56, p. 1–5.

Drooger, C. W., 1969, Voltooid verleden tija, heden en toekomst (Past, present, and future): Inaugural dissertation, Univ. of Utrecht, Schotanus & Jons, Utrecht, p. 1–16.

————, 1972, Late Neogene epoch boundaries and the classical European stratigraphy (abstr.): 24th Int. Geol. Cong. (Montreal), Abstr., p. 536.

————, 1974, The boundaries and limits of stratigraphy: Koninklijke Nederl. Akad. Wetenschappen, Amsterdam, Proc. Series B. v. 77, no. 3, p. 159–176.

———— **et al.,** 1964, Symposium on micropaleontological lineages and zones used for biostratigraphic subdivisions of the Neogene: Bern, 91 p. (multilith). (see Introduction, p. 1–3).

* **Drushchits, V. V.,** and **V. N. Shimansky,** 1962, On volume of Paleozoic era (Title translated from Russian): Dokladi Akad. Nauk SSSR, v. 144, no. 5, p. 1115–1118.

Dunbar, C. O., 1972, Stratigraphic boundaries and problems in their selection (abstr.): *in* The Age of the Dunkard, Symposium Abstr. and Reference Papers (I.C. White Memorial Symposium), p. 35–42, West Virginia Geol. Surv., Morgantown. (Reprinted from Round Table Conf. on Permo-Carboniferous Stratigraphic Problems, West Virginia Geol. Survey, 1951).

————, and **J. Rodgers,** 1957, Principles of stratigraphy: Wiley, New York, 356 p.

Dunn, P. R., K. A. Plumb, and **H. G. Roberts,** 1966, A proposal for time-stratigraphic subdivision of the Australian Precambrian: Geol. Soc. Australia Jour., v. 13, pt. 2, p. 593–608.

————, **B. P. Thomson,** and **K. Rankama,** 1971, Late Pre-Cambrian glaciation in Australia as a stratigraphic boundary: Nature, v. 231, no. 5304, p. 498–502.

Dyufur, M. S., 1971, Ob otnositel'nom (otnositel'no-stratigraficheskom) napravlenii v opredelenii soderzhaniya ponyatiya "fatsiya" (Relative (relative-stratigraphic) trend in defining the "facies" concept): Leningrad Univ. Vestn., Geol. Geogr., no. 24, p. 5–15. (Engl. summary).

Eames, F. E., et al., 1962, Fundamentals of mid-Tertiary stratigraphical correlation: Cambridge Univ. Press, 163 p.

Eaton, J. E., 1928, Divisions and duration of the Pleistocene in southern California: Am. Assoc. Petroleum Geol. Bull., v. 12, no. 2, p. 111–140.

————, 1929, The by-passing and discontinuous deposition of sedimentary materials: Am. Assoc. Petroleum Geol. Bull., v. 13, no. 7, p. 713–761.

————, 1931, Standards of correlation: Am. Assoc. Petroleum Geol. Bull., v. 15, no. 4, p. 367–385.

————, 1932, Time-equivalent versus lithologic extension of formations: Am. Assoc. Petroleum Geol. Bull., v. 16, no. 10, p. 1043–1044.

————, 1941, Reply to discussion by H. D. Hedberg of "Technique of stratigraphic nomenclature" by C. W. Tomlinson: Am. Assoc. Petroleum Geol. Bull., v. 25, no. 12, p. 2208–2210.

ECAFE Secretariat, 1970, Documents relating to organization of a continuing working group and standards for stratigraphic correlation in the ECAFE region: *in* Stratigraphic correlation between sedimentary basins of the ECAFE region, U.N. Econ. Comm. Asia Far East, v. 2 of Mineral Resources Development Series no. 36, p. 1–15, United Nations, New York.

Eckel, E. C., 1901, The formation as the basis for geologic mapping: Jour. Geology, v. 9, p. 708–717.

Eeden, O. R. van, 1971, South African code of stratigraphic terminology and nomenclature: Geol. Soc. South Africa, Trans., v. 74, pt. 3, p. 111–131.

* **Egoïan, V. L.,** 1973, Stratotip i stratigraficheskaia granitsa (Stratotype and stratigraphic boundary): Izv. Akad. Nauk SSSR, ser. geol., no. 2.

Eicher, D. L., 1968, Geologic time: Foundations of Earth Science Series, Prentice-Hall, New York, 149 p. (See p. 95–116.)

Elias, M. K., 1945, Geological calendar (Indications of periodicity in nature and succession of geological periods): Am. Assoc. Petroleum Geol. Bull., v. 29, no. 7, p. 1035–1043.

* **Eliseev, V. I.,** 1965, Some remarks on principle of establishment and inner subdivision of the Quaternary system (Title translated from Russian): Dokladi Akad. Nauk, v. 161, no. 2, p. 413–416.

Ellenberger, F., 1972, Quelques remarques historiques sur "la maladie infantil" de la stratigraphie: *in* Colloque sur les méthodes et tendances de la stratigraphie (Orsay, 1970), BRGM France, Mém 77, pt. 1, p. 27–30.

Enay, R., 1962, La nomenclature stratigraphique du Jurassique terminal, ses problèmes et sa normalisation: BRGM France, Dept. Inform. Géol. Bull. Trimest., v. 15, no. 57, p. 1–9.

———, 1966, L'oxfordien dans la moitié sud Jura français - étude stratigraphique: v. 1, fasc. 8, Imprimerie A. Rey, Lyon, France, 310 p. (See particularly Le cadre stratigraphique, p. 36–42.)

——— (ed.), 1971, Problèmes de zonation de quelques étages du Jurassique en Europe (communications presented by various French geologists at Luxembourg Colloque): BRGM France, Mém. 75, (Colloque du Jurassique, Luxembourg, 1967), p. 511–512.

* **Epshteyn, S. V.,** 1961, On the question of principles and methods of stratigraphic subdivision of the Quaternary system (Title translated from Russian): Mat. VSEGEI, vypusk 42, p. 19–36.

Erben, H. K., 1959, Fortschritte der Paläontologie im letzten Jahrzehnt: Naturwiss. Rundschau, Jahrg. 12, H. 4, p. 119–124.

———, 1961, Ergebnisse der 2. Arbeitstagung über die Silur/Devon-Grenze und die Stratigraphie von Silur und Devon: Bonn und Brüssel (1960): Deutsche Geol. Gesell. Zeit., Bd. 113, no. 1, p. 81–84.

——— (ed.), 1962, Internationale Arbeitstagung über die Silur/Devon-Grenze und die Stratigraphie von Silur und Devon: Bonn-Bruxelles (1960), Stuttgart, 315 p.

———, 1962, Diskussion zur Silur/Devon-Grenze: Symposium Silur/Devon-Grenze (1960), Stuttgart, p. 307–309.

———, 1972, Replies to opposing statements: Newsl. Strat. v. 2, no. 2, p. 79–95, Leiden.

Ericson, D. B., M. Ewing, and **G. Wollin,** 1963, Pliocene-Pleistocene boundary in deep-sea sediments: Science, v. 139, no. 3556, p. 727–737.

Evernden, J. F., and **R. K. S. Evernden,** 1970, The Cenozoic time scale; Geol. Soc. America Spec. Paper 124, p. 71–90.

Eysinga, F. W. B. van, 1970, Stratigraphic terminology and nomenclature; a guide for editors and authors: Earth-Science Reviews, v. 6, no. 4, p. 267–288.

Fairbridge, R. W., 1968, Quaternary Period: *in* Encyclopedia of geomorphology, Reinhold, New York, p. 912–931.

————, 1972, Fundamental considerations for an international agreement on the boundary between the Pleistocene and Holocene: Manuscript for INQUA Holocene Commission, 21 p.

Faul, H., 1960, Geologic time scale: Geol. Soc. America Bull., v. 71, no. 5, p. 637–644.

Fenton, C. L., and **M. A. Fenton,** 1928, Ecologic interpretation of some biostratigraphic terms, Part 1, Faunule and zonule: Am. Midland Naturalist, v. 11, no. 1, p. 1–23.

———— and **M. A. Fenton,** 1930, Ecologic interpretation of some biostratigraphic terms, Part 2, Zone, subzone, facies, phase: Am. Midland Naturalist, v. 12, no. 5, p. 145–153.

Fiege, K., 1926, Dei paläontoligischen Grundlagen der geologischen Zeitmessung: Naturwiss., Mh. v. 24, p. 77–91, Leipzig.

————, 1951, The zone, base of biostratigraphy: Am. Assoc. Petroleum Geol. Bull., v. 35, no. 12, p. 2582–2596.

————, 1952, Sedimentationszyklen und Epirogenese: Deutsche Geol. Gesell. Zeit., v. 103 (1951), p. 17–22.

————, 1969, Sedimentationzyklen als Zeitmarken: Deutsche Geol. Gesell. Zeit., v. 118 (1966), pt. 2, 260–265.

Fischer, A. G., 1969, Geological time-distance rates: The Bubnoff unit: Geol. Soc. America Bull., v. 80, no. 3, p. 549–551.

————, **H. E. Wheeler,** and **V. S. Mallory,** 1954, Arbitrary cut-off in stratigraphy: Discussion: Am. Assoc. Petroleum Geol. Bull., v. 38, no. 5, p. 926–931.

Fisher, D. W., 1956, Intricacy of applied stratigraphic nomenclature: Jour. Geology, v. 64, no. 6, p. 617–627.

Fisher, N. H., 1969, Steps taken for the standardization of stratigraphic nomenclature in Australia: *in* Stratigraphic correlation between sedimentary basins of the ECAFE region, Third symposium on development of petroleum resources Asia and Far East (Tokyo, 1969), U.N. Econ. Comm. Asia Far East, Mineral Resources Development Series, no. 30, p. 10–12.

Fisher, W. L., 1961, Stratigraphic names in the Midway and Wilcox Groups of the Gulf Coastal Plain: Gulf Coast Assoc. Geol. Soc. Trans., v. 11, p. 263–295.

Fleisher, R. L., 1974, Cenozoic planktonic foraminifera and biostratigraphy, Arabian Sea Deep Sea Drilling Project, Leg 23A: Initial reports of the Deep Sea Drilling Project, v. 23, p. 1001–1072, Washington.

Forgotson, J. M. Jr., 1957, Nature, usage and definition of marker-defined vertically segregated rock units: Geol. Notes, Am. Assoc. Petroleum Geol. Bull., v. 41, no. 9, p. 2108–2113.

————, 1957, Stratigraphy of the Comanchean Cretaceous Trinity Group: Am. Assoc. Petroleum Geol. Bull., v. 41, no. 10, p. 2328–2363.

France, Comité Français de Stratigraphie, 1960, Principes de classification et de nomenclature stratigraphique (ed. H. Tintant), (multilith copies), 8 p.

————, 1962, Principes de classification et de nomenclature stratigraphiques: available from A. Blondeau, Géologie des Bassins Sédimentaires Univ., Paris VI, 4 place

Jussieu, 75005, Paris, 15 p. Translations by F. de Rivero: Bol. Inform. Assoc. Venezolana Geol. Min. y Petroleo, v. 8, no. 8, (Spanish) p. 224–237; (English) p. 238–250.

Frank, M., 1938, Zur Frage der Richtprofile: Deutsche Geol. Gesell. Zeit., v. 90. p. 227–230.

Franke, D., 1962, Zu Fragen geologischer Terminologie und Klassifikation: (1) Der Begriff Formation: Zeit., Angew. Geol., v. 8, no. 4, (Berlin), p. 208–214.

——, 1963, Zu Fragen geologischer Terminologie und Klassifikation: (II) Der Begriff Fazies: (1. Teil), Zeit. Angew. Geol., H. 1, p. 39–45; (2. Teil), Zeit. Angew. Geol., H. 2, p. 97–102; (3. Teil), Zeit. Angew. Geol., H. 3, p. 153–157.

Frarey, M. J., and W. F. Fahrig, 1972, Subdivision of Precambrian: an interim scheme to be used by the U.S. Geological Survey: Discussion: Am. Assoc. Petroleum Geol. Bull., v. 56, no. 10, p. 2083–2084.

Frebold, H., 1924, Ammonitenzonen und Sedimentationszyklen in ihrer Beziehung zueinander: Centb. für Mineralogie, Geol. u. Pal., Jahrg. 1924, p. 313–320.

Frech, F., 1899, Über Abgrenzung und Benennung der geologischen Schichtengruppen: Int. Geol. Cong. (St. Petersburg, 1897), p. 27–52.

Frye, J. C., 1968, Development of Pleistocene stratigraphy in Illinois: in The Quaternary of Illinois, Univ. of Illinois, College of Agriculture Spec. Publ. 14, p. 3–10.

—— and A. B. Leonard, 1953, Definition of time line separating a glacial and interglacial age in the Pleistocene: Am. Assoc. Petroleum Geol. Bull., v. 37, no. 11, p. 2581–2586.

—— and G. M. Richmond, 1958, Note 20 (of Am. Comm. Strat. Nomen.)—Problems in applying standard stratigraphic practice in nonmarine Quaternary deposits: Am. Assoc. Petroleum Geol. Bull., v. 42, no. 8, p. 1979–1983.

—— and H. B. Willman, 1962, Note 27 (of Am. Comm. Strat. Nomen.)—Morphostratigraphic units in Pleistocene stratigraphy: Am. Assoc. Petroleum Geol. Bull., v. 46, no. 1, p. 112–113.

Füchsel, G. C., 1761, Historia terrae et maris, ex historia Thuringiae per montium descriptionem: Actorum Academiae Electoralis Moguntinae Scientarum Utilium, quae Erfordiae Est. t. 2, p. 44–208.

Furon, R., 1960, Infracambrian: Lexique Strat. Int., v. 8, Termes Stratigraphiques Majeurs, C.N.R.S., Paris, 74 p.

Furrer, H., 1939, Geologische Untersuchungen in der Wildstrubel-Gruppe: Thesis, Sonderabdruck aus den Mitteilungen der Naturforschenden Gesell. in Bern aus dem Jahre 1938, p. 35–167.

Gabilly, J., 1971, Méthodes et modeles en stratigraphie du Jurassique: BRGM France, Mém. 75 (Colloque du Jurassique, Luxembourg, 1967), p. 5–16.

Gage, M., 1966, Geological divisions of time: New Zealand Jour. Geol. Geophys., v. 9, no. 4, p. 399–407.

* Ganeshin, G. S., et al., 1961, Volume, contents and terminology of stratigraphic subdivisions of the Quaternary system (Title translated from Russian): Sov. Geol., no. 8, p. 3–15.

* Gar´kovets, V. G., et al., 1957, On the importance of facies paragenetic establishment of suites during facies—paleogeographical forecast and detailed prospecting surveying works in the Middle Asia (Title translated from Russian): Izv. Akad. Nauk Uzbekskoi SSR, ser. geol., no. 1, p. 13–16.

Garrett, P. A., 1960, Nomenclature of the Keuper Series: Nature, v. 187, no. 4740, p. 868–869.

Gartner, S. Jr., 1969, Correlation of Neogene planktonic foraminifer and calcareous nannofossil zones: Gulf Coast Assoc. Geol. Soc. Trans., v. 19, p. 585–599.

Gastil, G., 1960, The distribution of mineral dates in time and space: Am. Jour. Sci., v. 258, p. 1–35.

Gealy, E. L., E. L. Winterer, and **R. Moberly, Jr.,** 1971, Methods, conventions, and general observations: Initial reports of Deep Sea Drilling Project, v. 7, p. 9–26, Washington.

Geczy, B., 1964, Szint, Életszint, Időszint (Zone, biozone, chronozone): Földtani Közlöny, v. 94, p. 132–135, Budapest.

———, 1971, Biozones et chronozones dans le Jurassique de Csernye (Montagne Bakony): BRGM France, Mém. 75 (Colloque du Jurassique, Luxembourg, 1967), p. 411–422.

Geikie, A., 1885, Text-book of geology: 2nd ed., Macmillan, London, 992 p. (See p. 626–631.)

* **Gekker, R. F.,** 1956, On the question of methods of biostratigraphy (Title translated from Russian): Geol. Sbornik, no. 2–3, Lvov, p. 137–157.

George, T. N., 1956, Biospecies, chronospecies and morphospecies: Systematics Association Publication no. 2, p. 123–137.

———, 1960, Fossils in evolutionary perspective: Science Progress, v. 48, no. 189, 28 p.

———, 1965, Stratigraphical systems: Report on discussion in Kashmir: Geol. Soc. London Proc. no. 1624, p. 109–113.

——— **et al.,** 1967, The stratigraphical code—Report of the Stratigraphical Code Sub-Committee: Geol. Soc. London Proc. no. 1638, p. 75–87.

——— and **R. H. Wagner,** 1969, Report of the International Union of Geological Sciences Subcommission on Carboniferous Stratigraphy: C. R. 6e Cong. Carbonif. (Sheffield, 1967), v. 1, p. xlii–xiv.

——— **et al.,** 1969, Recommendations on stratigraphical usage: Geol. Soc. London, Proc. no. 1638, p. 139–166. (2nd revision of 1967 Report of Stratigraphical Code Sub-Committee.)

——— and **R. H. Wagner,** 1970, Report and procedures of the meeting of the IUGS Subcommission on Carboniferous Stratigraphy held in Liège, 16th to 18th April, 1969: Colloque sur la Stratigraphie du Carbonifère, Univ. de Liège, v. 55, p. 151–165.

——— and **R. H. Wagner,** 1972, Int. Union Geol. Sci., Subcom. on Carboniferous Stratigraphy, Proceedings and Report of the General Assembly at Krefeld, August 21–22, 1971: C.R. 7me Congrès Carbonifère, Krefeld (1971), v. I, p. 139–147.

* **Gerasimov, I. P.,** 1958, Problems of the Quaternary period. Results of the All Union meeting Moscow, May 1957 (Title translated from Russian): Priroda, no. 4, p. 49–54.

Gerasimov, N. P., 1937, Uralian Stage of the Permian (Title translated from Russian): Kazan Univ. Scientific Memoirs, v. 97, kn 3–4, Geol. no. 8–9, p. 3–68 (English summary, p. 65–68).

Geyer, O. F., 1973, Grundzüge der Stratigraphie und Fazieskunde, v. 1, Stuttgart, 279 p. (See particularly III Stratigraphie und Geochronologie, p. 177–259.)

Gignoux, M., 1926, Géologie stratigraphique: Masson, Paris, 588 p.

———, 1936, Géologie stratigraphique: Masson, Paris, 709 p.

————, 1952, La notion de temps en géologie et la tectonique d'écoulement par gravité: 18th Int. Geol. Cong. (London), pt. 13, p. 90–96.

————, 1955, Stratigraphic geology (English translation by G. G. Woodford of the 4th French edition, 1950, of Géologie Stratigraphique): W. H. Freeman & Co., San Francisco, Calif., 682 p.

————, 1960, Géologie stratigraphique: 5ᵉ édit., Masson, Paris, 759 p.

Gilluly, J., 1949, Distribution of mountain building in geologic time: Geol. Soc. America Bull., v. 60, p. 561–590.

Ginsburg, L., 1972, Rapport sur les vertébrés. L'apport de la paléontologie des vertébrés à la stratigraphie: *in* Colloque sur les méthodes et tendances de la stratigraphie (Orsay, 1970), BRGM France, Mém. 77, pt. 1, p. 339–344.

Gintsinger, A. B., and **M. K. Vinkman,** 1969, K voprosu o vozrastnoy migratsii granits svit (Problems of nonisochronous boundaries between various series): *in* Problemy stratigrafii, Sib. Nauch.-Issled. Inst. Geol. Geofiz. Miner. Syr'ya, Tr., no. 94, p. 107–111.

* **Gladenkov, Yu. B.,** 1972, Nekotorye diskussionye voprosy stratigrafii (Some controversial problems in stratigraphy): Izv. AN SSSR, ser. geol., no. II.

Glaessner, M. F., 1945, Principles of micropaleontology: Melbourne Univ. Press, Australia, 296 p. (See particularly p. 213–226.)

————, 1953, Time-stratigraphy and the Miocene epoch: Geol. Soc. America Bull., v. 64, no. 6, p. 647–658.

————, 1954, Time-stratigraphy of the late Pre-Cambrian: Pan Indian Ocean Sci. Cong. Proc., sec. C, p. 66–68.

————, 1960, West-Pacific stratigraphic correlation: Nature, v. 186, no. 4730, p. 1039–1040.

————, 1963, Preliminary report on generalized stratigraphic correlation between sedimentary basins in the ECAFE area: 2nd symposium on the development of petroleum resources of Asia and the Far East, U.N. Econ. Comm. Asia Far East, Mineral Resources Development Series, Proc., v. 1, no. 18, p. 139–144.

————, 1963, The base of the Cambrian: Geol. Soc. Australia Jour., v. 10, pt. 1, p. 223–241.

————, 1963, The dating of the base of the Cambrian: Geol. Soc. India Jour., v. 4, p. 1–11.

————, 1970, Notes concerning a chronostratigraphic scale for the ECAFE region: *in* Stratigraphic correlation between sedimentary basins of the ECAFE region, U.N. Econ. Comm. Asia Far East, v. 2 of Mineral Resources Development Series no. 36, p. 25–28, United Nations, New York.

———— et al., 1948, Stratigraphical nomenclature in Australia: Australian Jour. Sci., v. 11, no. 1, p. 7–9.

Glass, B., et al., 1967, Geomagnetic reversals and Pleistocene chronology: Nature, v. 216, November 4, 1967, p. 437–442.

Glumicić-Holland, N., and **Z. Boškov-Štajner,** 1967, Stratigrafska klasifikacija i terminologija (Stratigraphic classification and terminology, Copenhagen 1961): Nafta, v. 18, nos. 3–4, p. 95–111.

Gol'bert, A. V., 1969, O litologicheskoi obosoblennosti stratigraficheskikh podrazdelenii i vyrazhenii ikh granits v geologicheskom razreze (Lithologic separation of stratigraphic units and the definition of their boundaries in the geologic section): *in*

Problemy Stratigrafi, Trudy SNIGGIMSa, no. 94, p. 112–120, Novosibirsk. (English transl. by Israel Program for Scientific Translations *in* Classification in Stratigraphy, publ. Jerusalem, 1971, for U.S. Dept. Int., and Nat. Sci. Foundation, p. 99–106.)

Goldich, S. S. et al., 1961, The Precambrian geology and geochronology of Minnesota: Minn. Geol. Surv. Bull., no. 41, 193 p. (See p. 150–168.)

Gordon, W. A., 1962, Problems of paleontological correlation with particular reference to Tertiary: Am. Assoc. Petroleum. Geol. Bull., v. 46, no. 3, p. 394–398.

Gorsel, J. T. van, 1973, The type Campanian and the Campanian-Maastrichtian boundary in Europe: Geol. en Mijnb., v. 52, no. 3, p. 141–146.

Gorsky, I. I., 1957, Biostratigraphy and geochronology of continental deposits (Title translated from Russian): Izv. Akad Nauk SSSR, ser. geol., no. 12, p. 33–46.

———, and **V. V. Menner,** 1963, Stratigraficheskaya komissiya na XXI sessii Mezhdunarodnogo geologicheskogo kongressa (The Stratigraphic Commission of the 21st Session of the International Geologic Congress): Problemy geologii na XXI sessii MGK. Moskva, Izd. Akad. Nauk SSSR, p. 40–51.

Gowda, S. S., 1970, An approach to the problem of stratigraphic taxonomy and nomenclature in India: Jour. Mines, Metals and Fuels (Calcutta), v. 18, no. 1, p. 3–4, 18.

Grabau, A. W., 1913, Principles of stratigraphy: A.G. Seiler and Co., New York, 1185 p. (See particularly p. 1097–1150.)

Gray, H. H., 1955, Thickness of bedding and parting in sedimentary rocks: Geol. Soc. America Bull., v. 66, no. 1, p. 147–148.

———, 1958, Definition of term formation in stratigraphic sense: Discussion: Am. Assoc. Petroleum Geol. Bull., v. 42, no. 2, p. 451–452.

Green, R., 1962, Zonal relationships in Lower Mississippian rocks of Alberta: Alberta Soc. Petroleum Geol. Jour., v. 10, no. 6, p. 292–307.

Greenough, G. B., 1819, A critical examination of the first principles of geology: Strahan and Spottiswoode, London, 336 p.

Gregory, J. W., and **B. H. Barrett,** 1913, General stratigraphy: Methuen & Co., London, 283 p.

Gressly, A., 1838, Observations géologiques sur le Jura soleurois: Soc. Helv. Sci. Nat. (Neuchatel), Nouv. Mém., v. 2, 349 p. (See particularly p. 8–26.)

Griffiths, J. C., 1949, Sedimentary facies in geologic history (Discussion): Geol. Soc. America Mem. 39, p. 140–141.

Gromov, V. I., et al., 1960, Principles of a stratigraphic subdivision of the Quaternary (Anthropogen) System and its Lower boundary: 21st Int. Geol. Cong. (Norden), pt. 4, p. 7–26.

Gubler, Y., 1972, Stratigraphie et sédimentologie, Introduction et rapport de synthèse: *in* Colloque sur les méthodes et tendances de la stratigraphie (Orsay, 1970), BRGM France Mém. 77, pt. 2, p. 523–534.

Gurari, F. G., 1969, O pravilakh stratigraficheskoy klassifikatsii (Principles of stratigraphic classification): *in* Problemy Stratigrafii, Sib. Nauch.-Issled. Inst. Geol. Geofiz. Min. Syr'ya, Tr., no. 94, p. 66–78.

———, **I. I. Nesterov,** and **M. Ya Rudkevich,** 1962, On stratification of Mesozoic and Cenozoic deposits of the Western-Siberia lowland (Title translated from Russian): Geol. i Geofiz., no. 3, p. 3–10.

———, and **L. L. Khalfin,** 1966, Reforma pravil stratigraficheskoy klassifikatsii neobkhodima (Reform essential in rules of stratigraphic classification): Geol. i Geofiz., Akad. Nauk SSSR, Sib. Div., no. 4, p. 3–14. (English translation *in* Int. Geol. Rev., v. 8, no. 11, 1966, p. 1261–1269.)

————, and **L. L. Khalfin,** 1966, Reforma pravil stratipgraficheskoy klassifikatsii neobkhodima (Reform essential in rules of stratigraphic classification): Geol. i Geofiz., Akad. Nauk SSSR, Sib. Div., no. 4, p. 3–14. (English translation *in* Int. Geol. Rev., v. 8, no. 11, 1966, p. 1261–1269.)

Haefeli, C., W. Maync, H. J. Oertli, and **R. F. Rutsch,** 1965, Die Typus-Profile des Valanginien und Hauterivien: Ver. Schweiz. Petrol-Geol., u.-Ing. Bull., v. 31, no. 81, p. 41–75.

Hamaqui, M., and **M. Raub,** 1965, Biostratigraphy. Type sections of Cretaceous formations in the Jerusalem-Bet Shemesh area: Stratigraphic Sections, pub. no. 1, State of Israel Ministry of Development, Geol. Survey, p. 26–39.

Hammen, T. van der, 1957, Climatic periodicity and evolution of South American Maestrichtian and Tertiary floras: Bol. Geol., v. 5, no. 2, Bogota, p. 49–91.

————, 1965 (1964), Paläoklima, Stratigraphie und Evolution: Geol. Rundschau, v. 54, no. 1, p. 428–441.

Hancock, J. M., 1966, Theoretical and real stratigraphy: Correspondence, Geol. Mag., v. 103, no. 2, p. 179.

Hanlon, F. N., G. A. Joplin, and **L. C. Noakes,** 1952, Review of stratigraphic nomenclature, Mesozoic of the Cumberland Basin: Australian Jour. Sci., v. 14, no. 6, p. 179–182.

Harbaugh, J. W., 1968, Stratigraphy and geologic time: William C. Brown Company, Dubuque, Iowa, 113 p.

Harland, W. B., 1967, Review of "Stratigraphy. An introduction to principles" by D. T. Donovan: Geol. Mag., v. 104, no. 2, p. 192–194.

————, 1968, On the principle of a Late Pre-Cambrian stratigraphical standard scale: 23rd Int. Geol. Cong. (Prague), v. 4, p. 253–264.

————, 1970, Time, space and rock (An essay on some fundamentals of stratigraphy): West Commemoration Volume, Printed at Today & Tomorrow's Printers & Publishers, Paridabad, India, p. 17–42.

————, 1971, Introduction to the Phanerozoic time-scale (A supplement): Geol. Soc. London, Special Pub. no. 5, p. 3–7.

————, 1973, Stratigraphic classification, terminology and usage—essay review of: An International Guide to Stratigraphic Classification, terminology and usage (H.D. Hedberg, ed.): Geol. Mag., v. 110, no. 6, p. 567–574.

————, 1974, The Pre-Cambrian Cambrian boundary: *in* Cambrian of the British Isles, Norden and Spitsbergen (ed. C.H. Holland), v. 2, Lower Palaeozoic rocks of the world, John Wiley, London, p. 15–42.

————, 1975, The two geological time scales: Nature, v. 253, p. 505–507.

———— **et al.,** 1972, A concise guide to stratigraphical procedure: Geol. Soc. London Quart. Jour., v. 128, p. 295–305.

Harpum, J. R., 1960, The concept of the geological cycle and its application to problems of Precambrian geology: 21st Int. Geol. Cong. (Norden), pt. 9, p. 201–206.

Harrington, H. J., 1965, Space, things, time and events—an essay on stratigraphy: Am. Assoc. Petroleum Geol. Bull., v. 49, no. 10, p. 1601–1646.

Hartono, H. M. S., 1970, Steps toward standardization of stratigraphic classification in Indonesia: *in* Stratigraphic Correlation between Sedimentary Basins of the ECAFE region, v. 2, U.N. Econ. Comm. Asia Far East, Mineral Resources Development Series, no. 36, p. 130–134.

Haug, E., 1911, Traité de géologie, II Les périodes géologiques: Colin, Paris, p. 539–928. (See particularly p. 539–564, Principes généraux de la stratigraphie.)

Hay, W. W., 1972, Probabilistic stratigraphy: Eclog. Geol. Helv., v. 65, no. 2, p. 255–266.

———, 1974, Implications of probabilistic stratigraphy for chronostratigraphy: *in* Contributions to the geology and paleobiology of the Caribbean and adjacent areas: Kugler Festschrift Volume, Verhandl. Naturf. Ges. Basel, v. 84, no. 1, p. 164–171.

——— and **P. Cepek,** 1969, Nannofossils, probability, and biostratigraphic resolution (abstr.): Am. Assoc. Petroleum Geol. Bull., v. 53, no. 3, p. 721.

Hayami, I., 1973, An evolutionary interpretation of biostratigraphic zones: Geol. Soc. Japan Jour., v. 79, no. 3, p. 219–235. (In Japanese with abstr. in English.)

——— and **T. Ozawa,** 1975, Evolutionary models of lineage-zones: Lethaia, v. 8, no. 1, p. 1–14.

Hays, J. D., 1971, Faunal extinction and reversals of the earth's magnetic field: Geol. Soc. America Bull., v. 82, no. 9, p. 2433–2447.

——— and **N. D. Opdyke,** 1967, Antarctic radiolaria, magnetic reversals, and climatic change: Science, v. 158, no. 3804, p. 1001–1011.

——— **et al.,** 1969, Pliocene-Pleistocene sediments of the equatorial Pacific: their paleomagnetic, biostratigraphic, and climatic record: Geol. Soc. America Bull., v. 80, no. 8, p. 1481–1514.

——— and **W. A. Berggren,** 1971, Quaternary boundaries and correlations: *in* The micropaleontology of the oceans, Cambridge Univ. Press, p. 669–691.

Hazel, J. E., 1970, Binary coefficients and clustering in biostratigraphy: Geol. Soc. America Bull., v. 81, no. 11, p. 3237–3252.

Hébert, E., 1881, Nomenclature et classification géologiques: Rapport du Comité Français, Commission pour l'unification de la nomenclature géologique, Ann. Sci. Géol., v. 11, no. 10, art. 4, p. 1–15.

Hedberg, H. D., 1937, Stratigraphy of the Rio Querecual section of Northeastern Venezuela: Geol. Soc. America Bull., v. 48, p. 1971–2024; (section entitled "Some Stratigraphic Principles", p. 1975–1977.)

———, 1941, Discussion of "Technique of stratigraphic nomenclature" by C. W. Tomlinson: Am. Assoc. Petroleum Geol. Bull., v. 25, no. 12, p. 2202–2206.

———, 1948, Time-stratigraphic classification of sedimentary rocks: Geol. Soc. America Bull., v. 59, no. 5, p. 447–462.

———, 1951, Nature of time-stratigraphic units and geologic time units: Am. Assoc. Petroleum Geol. Bull., v. 35, no. 5, p. 1077–1081.

———, 1954, Procedure and terminology in stratigraphic classification: 19th Int. Geol. Cong. (Algiers), fasc. 13, p. 205–233.

———, 1958, Stratigraphic classification and terminology: Am. Assoc. Petroleum Geol. Bull., v. 42, no. 8, p. 1881–1896.

———, 1959, Stratigraphic classification and terminology: Alberta Soc. Petroleum Geol. Jour., v. 6, no. 8, p. 192–208.

———, 1959, Towards harmony in stratigraphic classification: Am. Jour. Sci., v. 257, p. 674–683.

————, 1961, The stratigraphic panorama (an inquiry into the bases for age determination and age classification of the earth's rock strata): Geol. Soc. America Bull., v. 72, no. 4, p. 499–518.

————, 1961, Stratigraphic classification of coals and coal-bearing sediments: Geol. Soc. America Bull., v. 72, no. 7, p. 1081–1088.

————, 1962, Les zones stratigraphiques — Remarques sur un article de P. Hupé: BRGM France Serv. Inform. Géol. Bull. Trimest., 14th year, no. 54, p. 6–11.

————, 1962, How to standardize stratigraphic language: World Oil, v. 154, April, p. 100–103.

————, 1965, Earth history and the record of the rocks: Am. Philosophical Soc. Proc., v. 109, no. 2, p. 99–104.

————, 1965, Chronostratigraphy and biostratigraphy: Geol. Mag., v. 102, no. 5, p. 451–461.

————, 1966, Note 33 (of Am. Comm. Strat. Nomen.)—Application to American Commission on Stratigraphic Nomenclature for amendments to Articles 29, 31, and 37 to provide for recognition of erathem, substage, and chronozone as time-stratigraphic terms in the Code of Stratigraphic Nomenclature: Am. Assoc. Petroleum Geol. Bull., v. 50, no. 3, p. 560–561.

————, 1967, Geochronology (stratigraphic): in International Dictionary of Geophysics (ed. by S.K. Runcorn), v. 1, Pergamon Press, p. 561–567.

————, 1967, Geologic periods and systems: in International Dictionary of Geophysics (ed. by S.K. Runcorn), v. 1, Pergamon Press, p. 600–605.

————, 1967, Status of stratigraphic classification and terminology: IUGS Geol. Newsletter, v. 1967, no. 3, p. 16–29, Antwerp.

————, 1968, Comments with respect to the Precambrian part of the geochronologic scale: C.R., Edmonton meeting of IUGS Commission on Geochronology, June 17, 1967, Annexe 1, p. 21–25.

————, 1968, Some views on chronostratigraphic classification: Geol. Mag., v. 105, no. 2, p. 192–199.

————, 1969, The influence of Torbern Bergman (1735-1784) on stratigraphy: Acta Univ. Stockholmiensis, Stockholm Contributions in Geology, v. 20, no. 2, p. 19–47, Pub. by Swedish Acad. of Sci.

————, 1969, Suggestions regarding stratigraphic classification and terminology in the ECAFE region: in Stratigraphic correlation between sedimentary basins of the ECAFE region, Third Symposium on development of petroleum resources Asia and Far East, Tokyo (1965), U.N. Econ. Comm. Asia Far East, Mineral Resources Development Series, no. 30, p. 6–9.

————, 1970, Stratigraphic boundaries—a reply: Eclog. Geol. Helv., v. 63, no. 2, p. 673–684.

————, 1971, Recently published reports of the International Subcommission on Stratigraphic Classification: Newsl. Strat., v. 1, no. 4, p. 59–60, Leiden.

————, 1973, Impressions from a discussion of the ISSC International Stratigraphic Guide, Hannover, October 18, 1972: Newsl. Strat., v. 2, no. 4, p. 173–180.

————, 1973, Reaction to an attack by Professor H. K. Erben on the International Subcommission on Stratigraphic Classification and its philosophy: Newsl. Strat., v. 2, no. 4, p. 181–184.

————, 1974, Basis for chronostratigraphic classification of the Precambrian: Precambrian Research, v. 1, no. 3, p. 165–177.

Heide, S. van der, 1960, Report on the stratigraphic colloquium: C.R., 4th Cong. on Carboniferous, Heerlen, 15–20 September, 1958, p. 23.

—— and **W. H. Zagwijn,** 1967, Stratigraphical nomenclature of the Quaternary deposits in the Netherlands: Med. Geol. Stichting, N.S., no. 18, p. 23–29.

Heilprin, A., 1887, The classification of the Post-Cretaceous deposits: Acad. Nat. Sci., Philadelphia Proc., p. 314–322.

Heim, A., 1934, Stratigraphische Kondensation: Eclog. Geol. Helv., v. 27, no. 2, p. 372–383.

Heirtzler, J. R., et al., 1968, Marine magnetic anomalies, geomagnetic field reversals, and motions of the ocean floor and continents: Jour. Geophys. Research, v. 73, no. 6, p. 2119–2136.

Heiskanen, W. A., 1967, Geochronology (stratigraphic): *in* International Dictionary of Geophysics, v. 1, p. 561–567.

Henbest, L. G., 1952, Significance of evolutionary explosions for diastrophic division of earth history—Introduction to the Symposium: Jour. Paleontology, v. 26, no. 3, p. 299–318.

Hennig, E., 1943, Formfragen der historischen Geologie: N. Jahrb. für Min. Geol. u Pal., Monatsheft, Abt. B, H. 7, p. 169–174.

Henningsmoen, G., 1955, Om navn på stratigrafiske enheter: Norges Geol. Undersøkelse, no. 191, p. 5–17.

——, 1961, Remarks on stratigraphic classification: Norges Geol. Undersøkelse, no. 213, p. 62–92.

——, 1964, Zig-zag evolution: Norsk Geologisk Tidsskrift, v. 44, pt. 3, p. 341–352.

——, 1973, The Cambro-Ordovician boundary: Lethaia, v. 6, no. 4, p. 423–439.

Henson, F. R. S., 1944, Stratigraphic classification and nomenclature: Geol. Mag., v. 81, no. 4, p. 166–169.

Herzog, L. F., W. H. Pinson, Jr., and **R. F. Cormier,** 1958, Sediment age determination by Rb/Sr analysis of glauconite: Am. Assoc. Petroleum Geol. Bull., v. 42, no. 4, p. 717–733.

Hinte, J. E. van, 1965, The type Campanian and its planktonic foraminifera: Koninkl. Nederl. Akad. van Wetenschappen, Amsterdam, Proc. ser., B, v. 68, no. 1, p. 8–28.

——, 1969, A *Globotruncana* zonation of the Senonian Subseries: 1st Int. Conf. on Planktonic Microfossils, Geneva (1967), Proc., v. 2, p. 257–266.

——, 1968, On the Stage: Geol. en Mijnbouw, v. 47, no. 5, p. 311–315.

——, 1969, The nature of biostratigraphic zones: 1st Int. Conf. on Planktonic Microfossils, Geneva (1967), Proc., v. 2, p. 267–272.

——, 1972, The Cretaceous time scale and planktonic-foraminiferal zones: Koninkl. Nederl. Akad. van Wetenschappen, Amsterdam, Proc., ser. B, v. 75, no. 1, 8 p.

Hölder, H., 1960, Geologie und Paläontologie in Texten und ihrer Geschichte: Freiberg/München, 565 p. (See p. 439–446.)

——, 1964, Jura: Handbuch der Stratigraphischen Geologie, v. 4, F. Enke, Stuttgart, 603 p. (See p. 1–10.)

——, 1971, Grundsätzliches zur Jura-Gliederung: BRGM France, Mém. 75 (Colloque du Jurassique, Luxembourg, 1967), p. 69–74.

—— and **A. Zeiss,** 1972, Zu der gegenwärtigen Diskussion über Prinzipien und

Methoden der Stratigraphie: Neues Jahrb. Geol. Palaeont., Monatsh., Jg. 1972, H. 7, p. 385–399, Stuttgart.

Holland, C. H., 1962, Diskussion zur Silur/Devon-Grenze: Symposium Silur/Devon-Grenze (1960), Stuttgart, p. 304.

———, 1964, Stratigraphical classification: Science Progress, v. 52, no. 207, p. 439–451.

———, 1965, The Siluro-Devonian boundary: Geol. Mag., v. 102, no. 3, p. 213–221.

———, **J. D. Lawson,** and **V. G. Walmsley,** 1962, Ludlovian classification—reply: Geol. Mag., v. 99, no. 5, p. 393–398.

———, **J. D. Lawson,** and **V. G. Walmsley,** 1963, The Silurian rocks of the Ludlow District, Shropshire: British Museum (Natural History) Bull., Geology, v. 8, no. 3, p. 95–171, London.

Holmes, A., 1947, The construction of a geological time-scale: Geol. Soc. Glasgow, Trans., v. 21, p. 117–152.

———, 1960, A revised geological time-scale: Edinburgh Geol. Soc., Trans., v. 17, pt. 3, p. 183–216.

———, 1963, Introduction: *in* The Precambrian (ed. by K. Rankama): Vol. 1 of the geologic systems, Interscience Publishers, New York, p. xi–xxiv.

Hornibrook, N. de B., 1965, A viewpoint on stages and zones: New Zealand Jour. Geol. Geophys., v. 8, no. 6, p. 1195–1212.

———, 1971, Inherent instability of biostratigraphic zonal schemes: New Zealand Jour. Geol. Geophys., v. 14, no. 4, p. 727–733.

Horusitzky, F., 1955, Geokronológiánk mai Problémái (On the problems of geochronology): Földtani Közlöny, v. 85, p. 106–121, Budapest.

Hottinger, L., and **H. Schaub,** 1960, Zur Stufeneinteilung des Paleocaens und des Eocaens: Eclog. Geol. Helv., v. 53, no. 1, p. 453–479.

———, **R. Lehmann,** and **H. Schaub,** 1964, Données actuelles sur la biostratigraphie du Nummulitique Méditerranéen: BRGM France, Mém. 28, pt. 2, Colloque sur le Paléogene, Bordeaux, September 1962, p. 611–652.

Howell, B. F., 1960, How should the Cambrian epochs and series be delimited?: 21st Int. Geol. Cong. (Norden), pt. 8, p. 37–39.

Hughes, N. F., 1964, Stages and boundaries in stratigraphy: *in* Colloque du Jurassique, Luxembourg (1962), Volume des Comptes Rendus et Mémoires publié par l'Institut grand-ducal, Section des Sciences naturelles, physiques et mathématiques, p. 30.

———, 1974, Beneficial regulation of procedure in editing stratigraphy: Lethaia, v. 7, no. 4, p. 283–286.

——— et al., 1967, A use of reference-points in stratigraphy: Geol. Mag., v. 104, no. 6, p. 634–635.

——— et al., 1968, Hierarchy in stratigraphical nomenclature (reply to discussion by P.C. Sylvester-Bradley): Geol. Mag., v. 105, no. 1, p. 79.

Hupé, P., 1960, Les zones stratigraphiques: BRGM France, Serv. Inform. Geol., Bull. Trimest., no. 49, p. 1–20.

Huxley, T., 1862, The anniversary address: Geol. Soc. London Quart. Jour., v. 18, p. xl–liv.

Ida, Kazuyoshi, 1958, Preliminary operation for the consecutive and mathematical analyses of stratigraphic succession: Geol. Survey of Japan, Bull., v. 9, no. 5, p. 301–314.

Imperial Oil Limited, Geological Staff, 1950, Devonian nomenclature in Edmonton area, Alberta, Canada: Am. Assoc. Petroleum Geol. Bull., v. 34, no. 9, p. 1807–1825.

India, Committee on Stratigraphic Nomenclature of, 1971, Code of stratigraphic nomenclature of India: India Geol. Survey Misc. Pub. no. 20, 28 p.

Indiana Geological Survey, Geological Names Committee, 1962, Application to Am. Comm. Strat. Nomen. for amendment on informal status of named aquifers, oil sands, coal beds, and quarry layers: Am. Assoc. Petroleum Geol. Bull., v. 46, no. 10, p. 1935.

Indonesia, Komisi sandi stratigraphi (ed. S. Martodjojo), 1973, Sandi stratigrafi Indonesia: Ikatan Geologi Indonesia, 19 p.

Indonesia, Commission for Stratigraphic Code of, 1975, Stratigraphic Code of Indonesia (revised edition) (ed. S. Martodjojo): 19 p., Bandung, Indonesia. (In Indonesian and in English.)

Ingle, J. C., 1973, Summary comments on Neogene biostratigraphy, physical stratigraphy, and paleo-oceanography in the maringal northeastern Pacific Ocean: Initial reports of the Deep Sea Drilling Project, v. 18, p. 949–960, Washington.

INQUA, Subcommission for European Quaternary Stratigraphy (Chairman, G. Lüttig), 1966, Die Subkommission für europäische Quartärstratigraphie der INQUA und ihre aktuellen Aufgaben: Eiszeitalter und Gegenwart, v. 17, p. 227–228.

International Geochronological Commission of I.U.G.S., 1967, Proposed recommendations for the world-wide geochronological scale: Int. Geol. Rev., v. 9, no. 3, p. 323–326.

International Geological Congress (Paris, 1878), 1880, C. R., 313 p. (See particularly p. 60–142.)

────── (Bologna, 1881), 1882, C. R., 657 p. (See particularly p. 89–126, 196–197, 429–658.)

────── (Berlin, 1885), 1888, C. R., 546 p. (See p. lxxiv–cxiv, 279–530.)

────── (London, 1888), 1891, C. R., 472 p. (See Appendices A, B, and C.)

────── (St. Petersburg, 1897), 1899, C. R., 464 p. (See p. cxlii–cli, 1–52.)

────── (Paris, 1900), 1901, C. R., v. 1, 637 p. (See p. 152–160, 192–203.)

International Subcommission on Stratigraphic Terminology (ISST), 1955, Circular-1, March 7, 1955, 55 p. (Organization and membership; questionnaire on general stratigraphic definitions and principles—10 topics.)

──────, 1956, Circular-2, February 14, 1956, 18 p. (Replies to questionnaire of Circular-1; plans for meeting at International Geological Congress in Mexico.)

──────, 1956, Circular-3, May 30, 1956, 6 p. Questionnaire on general stratigraphic principles—24 questions.)

──────, 1956, Circular-4, August 4, 1956, 2 p. (Preliminary results of questionnaire of Circular-3.)

──────, 1956, Circular-5, August 23, 1956, 23 p. (Replies to questionnaire of Circular-3.)

──────, 1958, Circular-6, April 25, 1958, 35 p. (Membership; Mexico City meeting; questionnaire of 62 questions on lithostratigraphy, biostratigraphy, chronostratigraphy; list of publications on stratigraphic classification and terminology.)

──────, 1959, Circular-7, March 20, 1959, 137 p. (Attachment 1: Replies to questions 1–25 of Circular-6 on general stratigraphic principles and lithostratigraphy, 46 p. Attachment 2: Replies to questions 26–38 of Circular-6 on biostratigraphy, 29 p.

Attachment 3: Replies to questions 39–62 of Circular-6 on chronostratigraphy and general stratigraphic principles, 41 p. Attachment 4: Draft statement on stratigraphic classification and terminology based on replies to questionnaire of Circular-6, 4 p. Attachment 5: Questionnaire on procedures—4 questions, 1 p. Attachment 6: List of publications on stratigraphic classification and terminology.)

————, 1960, Circular-8, February 1, 1960, 32 p. (Section 1: Replies to questionnaire of Att.-5 of Circular-7, 13 p. Section 2: Draft of statement of principles of stratigraphic classification and terminology, 6 p. Section 3: List of active members, 1 p. Section 4: List of publications on stratigraphic classification and terminology, 4 p. Section 5: Additional replies to questionnaire of Circular-6, 1 p.)

————, 1960, Circular-9, March 15, 1960, 18 p. (Statement of definitions of some stratigraphic terms to accompany statement of principles of stratigraphic classification and terminology.)

————, 1960, Circular-10, July 15, 1960, 63 p. (Statement of principles of stratigraphic classification and terminology, with appended glossary of terms; comments by members on statement of principles; comments by members on glossary of terms.)

————, 1961, Report 1, Principles of stratigraphic classification and terminology: 21st Int. Geol. Cong. (Norden) Proc., Part 25, 38 p. (Italian transl., 1963, Riv. Ital. Pal. Strat., v. 69, no. 3, p. 429–455.)

————, 1961, Circular-11, July 1, 1961, 60 p. (Report on Copenhagen meetings; current status and final draft of subcommission's statement on principles of stratigraphic classification and terminology, with accompanying glossary of terms; subjects for current discussion; publications on stratigraphic classification and terminology.)

————, 1962, Circular-12, May 15, 1962, 27 p. (Questionnaire on principles and procedure for definition of stratigraphic systems—25 questions; publications on stratigraphic classification and terminology.)

————, 1963, Circular-13, May 15, 1963, 70 p. (Replies to questionnaire of Circular-12; letters from Sigal, Pruvost, Roger, and von Gaertner.)

————, 1964, Circular-14, March 15, 1964, 15 p. (Draft of statement regarding definition of Systems.)

————, 1964, Circular-15, August 1, 1964, 53 p. (Procedure with regard to statement on definition of geologic systems; final draft of statement on definition of geologic systems; votes of members; comments of members.)

————, 1964, Report-2, Definition of geologic systems: 22nd Int. Geol. Cong. (India) Proc., Part 18, 26 p. (Reprinted in large part in Am. Assoc. Petroleum Geol. Bull., 1965, v. 49, no. 10, p. 1694–1703.)

————, 1965, Circular-16, June 1, 1965, 18 p. (Reports on New Delhi and Kashmir meetings; report on definition of geologic systems; membership and composition of subcommission; name of subcommission; questionnaire—7 questions; procedures.)

————, 1966, Circular-17, July 1, 1966, 33 p. (Comments on Circular-16 by members; membership and organization of subcommission; name of subcommission; commentaries by members on Statement of Definition of Geologic Systems; suggestions for inviting outside commentaries; work program of subcommission; plans for International Stratigraphic Code.)

International Subcommission on Stratigraphic Classification (ISSC), 1967, Circular-18, February 15, 1967, 44 p. (Invited outside commentaries on Report on Definition of Geologic Systems; commentaries from members; draft statement on lithostratigraphic units.)

————, 1967, Circular-19, December 28, 1967, 34 p. (Comments on draft report on lithostratigraphic units of Circular-18; other comments on Circular-18; meeting of Int. Com. on Strat. in Berne, May 1967; plans for meeting at Prague Congress; some new national stratigraphic codes; membership of subcommission; Russian publications on stratigraphic classification and terminology; other publications on stratigraphic classification and terminology.)

————, 1968, Circular-20, January 15, 1968, 27 p. (Draft report on stratotypes.)

————, 1968, Circular-21, June 15, 1968, 24 p. (Plans for Prague meeting; additions to membership; second draft of report on lithostratigraphy.)

————, 1968, Circular-22, July 10, 1968, 50 p. (Comments on draft report on lithostratigraphy of Circular-20; further comments on Circular-19; further comments on Report on Definition of Geologic Systems.)

————, 1969, Circular-23, July 5, 1969, 28 p. (Report on unofficial meeting of subcommission in Prague on August 23, 1968, during Int. Geol. Cong; introductory remarks by chairman at Prague meeting; status of preparation of chapters of proposed International Guide to Stratigraphic Classification and Usage; status of International Geological Correlations Program (IGCP); membership; library depositories; some recent literature references. Attachment: Extract from Unesco document on "Structure and content of an International Geological Correlation Program (IGCP)".)

————, 1969, Circular-24, July 10, 1969, 28 p. (Comments on second draft of report on lithostratigraphic units of Circular-21; third draft of report on lithostratigraphic units.)

————, 1969, Circular-25, July 15, 1969, 29 p. (Second draft of report on stratotypes; comments on Circular-20 received after July 1, 1968.)

————, 1970, Circular-26, January 20, 1970, 44 p. (Final draft of Preliminary Report on Lithostratigraphic Units; votes of members; dissenting comments or qualifying remarks; editorial comments of members on third draft of report on lithostratigraphic units of Circular-24.)

————, 1970, Circular-27, January 25, 1970, 60 p. (Final draft of Preliminary Report on Stratotypes; votes of members; dissenting comments or qualifying remarks; editorial comments of members on draft report on stratotypes of Circular-25.)

————, 1970, Report-3, Preliminary report on lithostratigraphic units: 24th Int. Geol. Cong. (Montreal), 30 p. (pre-Congress publication.)

————, 1970, Report-4, Preliminary report on stratotypes: 24th Int. Geol. Cong. (Montreal), 39 p. (pre-Congress publication.)

————, 1970, Circular-28, June 5, 1970, 36 p. (Membership; International Geological Correlation Program; publication of reports on lithostratigraphic units and on stratotypes; status of International Stratigraphic Guide; library depositories; recent stratigraphic codes; late comments on lithostratigraphic and stratotype reports; dissenting comments on ISSC philosophy by Dr. H. K. Erben; list of active members of subcommission.)

————, 1970, Circular-29, June 10, 1970, 48 p. (Preliminary report on biostratigraphic units.)

————, 1970, Circular-30, August 15, 1970, 80 p. (Preliminary report on chronostratigraphic units; glossary of chronostratigraphic terms.)

————, 1970, Circular-31, November 10, 1970, 58 p. (Request for additional comments on drafts on biostratigraphic and chronostratigraphic units; relations of subcommission to International Geological Correlations Program; responses to proposal of Professor Erben in Circular-28; membership; draft of an introductory chapter to the International Stratigraphic Guide; further comments on Circulars 26 and 27.)

———, 1971, Circular-32, February 1, 1971, 101 p. (Comments on first draft of report on biostratigraphic units of Circular-29; additional comments on proposal of Professor Erben in Circular-28.)

———, 1971, Circular-33, February 5, 1971, 79 p. (Second draft of preliminary report on biostratigraphic units.)

———, 1971, Circular-34, May 10, 1971, 93 p. (Second draft of preliminary report on chronostratigraphic units; comments on first draft of report on chronostratigraphic units of Circular-30.)

———, 1971, Circular-35, May 20, 1971, 28 p. (Comments on first draft of Introduction to Guide of Circular-31; further comments on Professor Erben's proposal in Circular-28; recent national stratigraphic codes and additions to the Lexicon of Stratigraphy; membership; memorandum from Professor Erben received February 4, 1971.)

———, 1971, Circular-36, June 15, 1971, 120 p. (Final draft of Preliminary Report on Biostratigraphic Units; votes; dissenting comments or qualifying remarks; comments received on second draft of report on biostratigraphic units of Circular-33.)

———, 1971, Report-5, Preliminary report on biostratigraphic units: 24th Int. Geol. Cong. (Montreal), 50 p. (pre-Congress publication.)

———, 1971, Report-6, Preliminary report on chronostratigraphic units: 24th Int. Geol. Cong. (Montreal), 39 p. (pre-Congress publication.)

———, Circular-37, December 3, 1971, 56 p. (Membership; plans for meetings at Montreal Congress; further responses to proposals of Professor Erben; late replies to Circulars 31, 33, 34, and 36; comment on Report-4 on stratotypes; recent national and regional stratigraphic codes and additions to Stratigraphic Lexicon; annual report of Subcommission.)

———, 1971, Circular-38, December 5, 1971, 56 p. (Draft of Introduction and Summary to an International Stratigraphic Guide.)

———, 1972, Circular-39, February 7, 1972, 85 p. (Responses to Circular-38; procedure for election of ISSC officers for 1972-1976; English translations of some recent USSR papers on stratigraphic classification. Attachment A: Final draft of Introduction and Summary to an International Stratigraphic Guide. Attachment B: Response sheet for nomination of officers.)

———, 1972, Circular-40, March 20, 1972, 11 p. (Nomination of officers of subcommission; suggestions for agenda of Montreal meetings; status of publication of Introduction and Summary; membership applications; late comments from members on previous circulars.)

———, 1972, Circular-41, May 3, 1972, 17 p. (Nomination of officers; distribution of ballots for voting. Appendix-A: Membership list.)

———, 1972, Circular-42, June 1, 1972, 5 p. (Agenda for meetings at Montreal Congress; membership applications; new national stratigraphic codes; correspondence from members; late responses to Circular-39.)

———, 1972, Report-7, (a) Introduction to an international guide to stratigraphic classification, terminology, and usage; (b) Summary of an international guide to stratigraphic classification, terminology, and usage; Lethaia, 1972, v. 5, no. 3, p. 283–323;

Boreas, 1972, v. 1, no. 3, p. 199–239. Spanish translation by C. Petzall, 1973, Bol. Geologia, v. 11, no. 22, p. 287–331, Caracas, Venezuela.

———, 1972, Circular-43, November 25, 1972, 59 p. (Report of meetings of Subcommission at Int. Geol. Cong. (Montreal), Aug. 22-24, 1972. Appendix-I: English text of "Main principles of draft of USSR stratigraphic code" by A.I. Zhamoida et al. Appendix-II: Texts of commentaries by *ex officio* members of Subcommission— McLaren, Ricour, Bolli, Pomerol, Gèorge, Maubeuge, Plumstead (by Clifford), Burwash, and Seneš. Appendix-III: Text of "Magnetic reversals" by Norman Watkins. Appendix-IV: Text of "Relation of lithostratigraphic units to chronostratigraphic units in deep sea sediments" by H.E. Cook (as presented by G. Cohee). Appendix-V: Text of "Application of stratigraphic principles (ISSC) in the work of the Jugoslav Oil Company" by Z. Boskov-Stajner. Appendix-VI: Text of letter of July 20, 1972, from I. Chlupač and statement by Czechoslovak Stratigraphic Commission on "trilingual paper" of Laffitte, Harland, Erben, et al., with reply of August 7 from Chairman of Subcommission. Appendix-VII: Letter of August 1 from I. Kobayashi on the term "arbet". Annual report of Subcommission. Letter of October 13, 1972 from A. Salvador.)

———, 1973, Circular-44, January 20, 1973, 22 p. (Membership procedure and applications; editorial board for Guide; proposed symposium on an international geochronological scale; comment on "trilingual paper" by Laffitte, Harland, Erben et al.; further discussions with Prof. Erben; comments on ISSC Report 7 from A. R. Lloyd and from Wilfred Walker; recent publications on stratigraphic classification.)

———, 1973, Circular-45, October 1, 1973, 55 p. (Membership applications and procedures; letters of comment on Circulars- 43 and -44; comments on preliminary published chapters of Guide; procedure for editorial revision of Guide; continuing discussion with Prof. Erben; library depositories; unconformity-bounded stratigraphic units; terminology of magnetostratigraphic units; proposed symposium on an international geochronological scale; annual report of Subcommission.)

———, 1974, Circular-46, February 25, 1974, 132 p. (Procedure used in preparing draft of bibliography for International Stratigraphic Guide; symposium on international geochronological scale; "Criteria for Deciding System Boundaries" by J.D. Lawson; unconformity-bounded units; magnetostratigraphic units; comments on Circular-45; relation of Subcommission to Int. Geol. Correlations Program; regional committee for ECAFE region; comments on some recent publications; replies to questions asked in Circular-45. Attachment-A: First draft of bibliography to accompany International Stratigraphic Guide.)

———, 1974, Circular-47, October 28, 1974, 86 p. (Discussion of preparation of draft of International Stratigraphic Guide and request for comments on draft. Attachment-A: October 1974 draft of International Stratigraphic Guide.)

Irving, E., 1971, Nomenclature in magnetic stratigraphy: Royal Astron. Soc., Geophys. Jour., v. 24, no. 5, p. 529–531.

———, 1972, Paleomagnetic stratigraphy: names or numbers?: Comments, Earth Sci.: Geophysics, v. 2, no. 4, p. 125–130.

———, 1974, Dissent magnetic—letter to editor: Geotimes, v. 19, no. 4, p. 14.

——— and **L. G. Parry,** 1963, The magnetism of some Permian rocks from New South Wales: Royal Astron. Soc., Geophys. Jour., v. 7, no. 4, p. 395–411.

Israel, Geol. Survey of Israel Stratigraphic Committee (Z. Reiss, chairman), 1960, Report no. 8, multilithed, 6 p.

————, 1961, Report no. 9: mimeographed sheets, 29 p.

Italy, Commissione Stratigrafica del Comitato Geologica d'Italia (A. Azzaroli, M. B. Cita, and R. Selli), 1968, Codice Italiano di Nomenclatura Stratigrafica (Italian Code of Stratigraphic Nomenclature): Servizio Geol. d'Italia Boll., v. 89, p. 3–22.

Ivanov, V. K., 1972, Omonimya terminu "gorizont" i pidrozdih Donets'koi nizhn'oi permi (Ambiguity of the term "horizon"; subdivision of the Lower Permian in the Donets Basin): Tektonika Strat., no. 3, p. 33–35 (with English summary).

* **Ivanova, E. A.,** 1955, On the question of connection of stages of evolution of organic world with the stages of evolution of the Earth's crust (Title translated from Russian): Dokladi Akad. Nauk SSSR, v. 105, N I, p. 154–157.

Jaanusson, V., 1960, The Viruan (Middle Ordovician) of Öland: Geol. Inst., Univ. Uppsala Bull., v. 38, p. 207–288.

James, G. A., and **J. G. Wynd,** 1965, Stratigraphic nomenclature of Iranian oil consortium agreement area: Am. Assoc. Petroleum Geol. Bull., v. 49, no. 12, p. 2182–2245.

James, H. L., 1960, Problems of stratigraphy and correlation of Precambrian rocks with particular reference to the Lake Superior region: Am. Jour. Sci. (Bradley volume), v. 258-A, p. 104–114.

————, 1972, Note 40 (of Am. Com. Strat. Nomen.)—Subdivision of Precambrian: An interim scheme to be used by U.S. Geological Survey: Am. Assoc. Petroleum Geol. Bull., v. 56, no. 6, p. 1128–1133.

————, 1972, Subdivision of Precambrian: Reply: Am. Assoc. Petroleum Geol. Bull., v. 56, no. 10, p. 2084–2086.

Japan, 1952, Code of stratigraphic nomenclature of the Geological Society of Japan: Geol. Soc. Japan Jour., v. 58, p. 112–113. (In Japanese, with stratigraphic unit terms in English.)

* **Jaworowski, K.,** 1964, Przestrzenna koncepcja facji i niektóre terminy facjalne: Przegl. Geol., no. 11, p. 461–463.

Jeffrey, C., 1973, Biological nomenclature: Special Topics in Biology Series for the Systematics Association, Edward Arnold, London, Publishers, 69 p.

Jeletzky, J. A., 1956, Paleontology, basis of practical geochronology: Am. Assoc. Petroleum Geol. Bull., v. 40, no. 4, p. 679–706.

————, 1965, Is it possible to quantify biochronological correlation?: Jour. Paleontology, v. 39, no. 1, p. 135–140.

Jenkins, D. G., 1965, Planktonic foraminiferal zones and new taxa from the Danian to Lower Miocene of New Zealand: New Zealand Jour. Geol. Geophys., v. 8, no. 6, p. 1088–1126.

————, 1966, Standard Cenozoic stratigraphic zonal scheme: Nature, v. 211, no. 5045, p. 178.

————, 1971, The reliability of some Cenozoic planktonic foraminiferal "datum-planes" used in biostratigraphic correlation: Jour. Foraminiferal Research, v. 1, no. 2, p. 82–86.

Jepsen, G. L., 1940, Paleocene faunas of the Polecat Bench Formation, Park County, Wyoming, Part 1: Am. Phil. Soc. Proc., v. 83, no. 2, p. 217–340. (See p. 219–243.)

Jewett, J. M., 1962, The concept of time in stratigraphic classification: Kansas Acad. Sci. Trans., v. 65, no. 2, p. 97–109.

Jukes-Brown, A. J., 1899, Zones and "chronological maps": Geol. Mag., Decade 4, v. 6, p. 216–219.

————, 1903, The term "Hemera": Geol. Mag., v. 10, p. 36–38.

Kadota, O., 1954, Notes on the usage of stratigraphic terms: Chiba Univ., Coll. Arts & Sci. Jour., v. 1, no. 3, p. 178–180.

Kaemmel, T. von, 1966, Das Äcquivalent der Biozone in der Radio-geochronologie— der Vertrauensbereich: Geologie, Jahrg. 15, H. 8, p. 989–992, Berlin.

Kahler, F., 1955, Stratigraphische Begriffe: Verh. Geol. Rundesanst, Wien., p. 242–246.

* **Kamysheva-Elpat'evskaya, V. G.,** 1960, On naming of paleontological zones (Title translated from Russian): Uchenaia zapiska Saratovskogo universiteta, v. 74, p. 5–6.

Kauffman, E. G., 1970, Population systematics, radiometrics and zonation—a new biostratigraphy: North Am. Paleont. Convention, September 1969, Proc. Part F, p. 612–666.

Kauter, K., F. Stammberger, and **G. Tischendorf,** 1968, Vergleichends Tabelle der letzten veröffentlichten geochronologischen Zeitskalen der phanerozoischen Epochen: IUGS Kommission für Geochronologie. Zeit. Angew. Geol., Bd. 14, H. 8, p. 440–442.

Kay, M., 1947, Analysis of stratigraphy: Am. Assoc. Petroleum Geol. Bull., v. 31, no. 1, p. 162–168.

————, 1956, Precambrian and Protozoic: Discussion: Am. Assoc. Petroleum Geol. Bull., v. 40, no. 7, p. 1722–1723.

————, 1957, Discussion of Note 17 (of Am. Com. Strat. Nomen.)—Suppression of homonymous and obsolete stratigraphic names: Am. Assoc. Petroleum Geol. Bull., v. 41, no. 8, p. 1890–1891.

* **Kazarinov, V. P.,** 1965, Theory and practical application of the lithological formation method (Title translated from Russian): Sov. Geol., no. 8, p. 54–68.

Kegel, W., 1937, Über Richtprofile: Deutsche Geol. Gesell. Zeit., Bd. 90, H. 4, p. 224–226.

* **Keller, B. M.,** 1950, Stratigraphic subdivisions (Title translated from Russian): Izv. Akad. Nauk SSSR, ser. geol., no. 6, p. 3–25.

* ————, 1964, Principles of establishment and subdivision of the Upper Pre-Cambrian (Title translated from Russian): *in* Stratigraphy of the USSR, v. 2, p. 578–586, Gosgeoltehizdat, Moskva.

————, 1973, Rifey i yego mesto v edinoy stratigraficheskoy shkale dokembriya (The Riphean and its place in the single stratigraphic scale of the Precambrian): Sov. Geol. no. 6, p. 3–17. (English translation *in* Int. Geol. Rev., v. 16, no. 6, p. 714–726.)

* ———— and **V. V. Menner,** 1955, The All-Union meeting on questions of stratigraphy (Title translated from Russian): Izv. Akad. Nauk SSSR, ser. geol., no. 4, p. 170–174.

———— and **V. V. Menner,** 1955, Conférence de l'Union sur les questions de stratig- raphie: Izv. Akad. Nauk SSSR, ser. geol., no. 4, (Fide Stepanov as translated to French by Mme. Jayet, S.I.G., Paris.)

———— et al., 1968, The main features of the Late Proterozoic paleogeography of the USSR: 23rd Int. Geol. Cong. (Prague) Proc., sec. 4, p. 189-202.

Keller, G., 1957, Fortschritte in der Methodik und Ergebnisse geologischer Zeit- rechnung: Naturwiss. Rundschau (Stuttgart), Jahrg. 10, H. 5, p. 169–172.

Kent, P. E., 1968, Comment on Hughes et al, 1967, Geol. Mag., v. 105, no. 1, p. 80.

Keyes, C. R., 1923, Contraposed criteria of geological classification: Pan-Am Geol., v. 39, no. 1, p. 51–54.

———, 1923, Uniformity in geological classification: Pan-Am Geol., v. 39, no. 3, p. 239–246.

———, 1923, Taxonomy of periods in geology: Pan-Am Geol., v. 40, no. 2, p. 151–156.

———, 1924, Global concurrence of geological periods of time: Pan-Am Geol., v. 41, no. 4, p. 317–320.

———, 1927, Standardization of geological terminology: Pan-Am Geol., v. 48, no. 3, p. 213–218.

———, 1931, Diastatic framework of our geological chronology: Pan-Am Geol., v. 56, no. 2, p. 85–115.

———, 1934, Practicability in geological nomenclature: Pan-Am Geol., v. 61, no. 3, p. 231–234.

———, 1935, Stratigraphic disuse of group rank: Pan-Am Geol., v. 63, no. 1, p. 73–76.

———, 1935, Priority vs. usage in geological terminology: Pan-Am Geol., v. 64, no. 2, p. 141–144.

———, 1936, Diastatic measure of biotic chronology: Pan-Am Geol., v. 66, no. 5, p. 363–376.

———, 1937, Homotaxial principle in geological classification: Pan-Am Geol., v. 67, no. 3, p. 215–230.

———, 1937, Absolute scale of geological ages: Pan-Am Geol., v. 67, no. 3, p. 231–234.

Khain, V. E., 1964, Main stages and some general regularities in the development of the earth's crust: 22nd Int. Geol. Cong. (India), Part 4, Proc., Sec. 4, p. 36–52.

——— and **A. B. Ronov,** 1960, World paleogeography and lithological associations of the Mesozoic Era: 21st Int. Geol. Cong. (Norden), Part 12, p. 152–164.

* **Khalfin, L. L.,** 1955, On some problems of regional stratigraphy (Title translated from Russian): Materialy Novosibirskoi konferencii po ucheniiu o geologicheskih faciiah, v. 1, p. 45–55, Novosibirsk.

———, 1959, Osadochnye geologicheskie formatsii v stratigraficheskom aspekte (The stratigraphic aspect of sedimentary geologic formations): Sov. Geol., no. 10.

* ———, 1960, Some notes about the results of the All-Union meeting on general problems of stratigraphic classification (Leningrad, January 17–22, 1955) (Title translated from Russian): Trudi Tomskogo Univ., v. 146, p. 110–116.

———, 1960, Printsip biostratigraficheskoi parallelizatsii (Principle of biostratigraphic parallelization): Trudy SNIGGIMSa, no. 8.

* ———, 1960, On tectonic stratigraphic trend in geology and on principles of stratigraphy (Title translated from Russian): in Osnyovnyi idei M.A. Usova v geologii, Izd. Akad. Nauk Kazahskoi SSR Alma-Ata, p. 381–393.

* ———, 1964, On M.A. Usov's view on origin of the main unit of regional stratigraphy (Title translated from Russian): Materialy po geologii i poleznym iskopaemym Zapadnoi Sibiri, p. 21–25, Tomsk.

———, 1967, Printsip posledovatel·nosti obrazovaniya geologicheskikh tel (printsip Stenona). Pravilo posledovatel·nosti naplastovaniya (pravilo Stenona-Khettona) (The The rule of subsequent bedding (The Steno-Hutton Rule): Trudy SNIGGIMSa, no. 57.

———, 1969, Printsip Nikitina-Chernysheva—teoreticheskaya osnova stratigraficheskoy

klassifikatsii (The principle of Nikitin-Chernyshev—the theoretical basis of strati-
graphic classification: *in* Problemy Stratigrafii, Trudy SNIGGIMSa, no. 94, p. 7–42,
Novosibirsk. English translation by Israel Program for Scientific Translations, *in*
Classification in Stratigraphy, pub. Jerusalem, 1971, for U.S. Dept. of Int. and Nat. Sci.
Foundation, p. 3–34.

——— (ed.), 1969, Classification in stratigraphy: Trudy SNIGGIMS Ministerstva Geol.
SSSR, no. 94, Series Stratigraphy and Paleontology. Translated from Russian to
English by Israel program for Scientific Translations, pub. Jerusalem, 1971 for U.S.
Dept. of Int. and Nat. Sci. Foundation, 178 p.

———, 1970, Printsip A.P. Karpinskogo i granitsy podrazdeleniy mezhdunarodnoy
stratigraficheskoy shkaly (MSSN) (The A.P. Karpinskij principle and the unit boundaries of the International Stratigraphic Time Scale): *in* Materialy po regional·noy
geologii Sibiri, Sib. Nauch.-Issled. Inst. Geol. Geofiz. Min. Syr·ya, Tr. no. 110, p.
4–10.

* **Khomisury, P. I.,** 1960, On the question of regulation of stratigraphic terminology
(Title translated from Russian): Sov. Geol., no. 3, p. 133–135.

Khramov, A. N., 1960, Paleomagnetism and stratigraphic correlation: (Published in
Russian by Gostoptechizdat, 1958) (English translation by A.J. Loikine), Canberra,
Australia, 178 p.

Kitts, D. B., 1966, Geologic time: Jour. Geology, v. 74, no. 2, p. 127–146.

Kleinpell, R. M., 1934, Difficulty of using cartographic terminology in historical geology:
Am. Assoc. Petroleum Geol. Bull., v. 18, no. 3, p. 374–379.

———, 1938, Miocene stratigraphy of California: Am. Assoc. Petroleum Geol., Tulsa,
Oklahoma, 450 p. (See particularly p. 87–136.)

——— (reported by J.S. Brown), 1960, Principles of biostratigraphy: Alberta Soc. Petroleum Geol. Jour., v. 8, no. 4, p. 136–137.

———, 1972 (?), Some of the historical context in which a micropaleontological stage
classification of the Pacific Coast middle Tertiary has developed: Pacific Section, Soc.
Ec. Pal. Min., Miocene Symposium volume, p. 89–110 (Mimeographed).

Knopf, A., 1949, Time in earth history: *in* G.L. Jepsen et al., Genetic Paleontology, and
Evolution, Part I, Geological Time, Princeton Univ. Press, p. 1–9.

Kobayashi, T., 1944, Concept of time in geology, Pt. 1, On the major classification of the
geological age: Imp. Acad. Tokyo Proc., v. 20, no. 7, p. 475–478.

Pt. 2, 1944, The length of the Sinian time estimated by the stratigraphical method: Ibid,
v. 20, no. 7, p. 479–498. Pt. 3, 1944, An instant in the Phanerozoic Eon and its
bearing on geology and biology: Ibid, v. 20, no. 10, p. 742–750. Pt. 4, 1945, An
explanation of the relation between mutation and saltation together with an advice to
the Uniformitarian: Ibid, v. 21, no. 1, p. 70–73. Pt. 5, 1945, Time scale of the
Diluvium and relation among various kinds of time in historical sciences: Ibid, v. 21,
p. 74–77. Pt. 6, 1958, Continuity among various kinds of time in geology: Ibid, ??

———, 1971, International Subcommission on Stratigraphic Classification of International Commission on Stratigraphy, IUGS: Jour. Geography (Tokyo Geographic Soc.), v. 80, no. 2, p. 126–128.

Koenig, J. W., 1961, Stratigraphic principles and policy: *in* The stratigraphic succession
in Missouri (ed. W. B. Howe and J. W. Koenig), State of Missouri, Geological Survey
and Water Resources, 2nd ser., v. 40, Rolla, Mo., 185 p. (See p. 137–158.)

Kölbel, H., 1963, Internationale Beschlusse zur stratigraphischen Gliederung und Nomenklature des Jura-Systems: Deutsche Geol. Gesell. Ber., v. 8, no. 4, p. 390–394.

Kopek, G., E. Dudich, Jr., and **T. Kecskeméti,** 1971, Le problème de coupes-repères, problème central des recherches stratigraphiques: Hung., Magy, Allami Földt. Intez., Evk., v. 54, no. 4, Pt. 1, p. 347–357 (with discussion).

———, **E. Dudich, Jr.,** and **T. Kecskeméti,** 1971, Opornyy razrez kak osnovnoy vopros stratigraficheskikh issledovaniy (Type section as basic problem in stratigraphic studies): Hung. Magy. Allami Földt. Intez., Evk., v. 54 (1969), no. 4, Pt. 2, p. 119–123.

Kosanke, R. M., et al., 1960, Classification of the Pennsylvanian strata of Illinois: Report of Investigations 214, Ill. State Geol. Survey, 84 p.

Kosygin, Y. A., and **V. A. Solov'yev,** 1969, Geologicheskiye formatsii i tektonika (Geological formations and tectonics): Geol. i Geofiz., no. 3, p. 17–24. (English translation *in* Int. Geol. Rev., v. 12, no. 7, 1970, p. 856–861.)

Kosygin, Yu. A., Yu S. Salin, and **V. A. Solovyev,** 1974, Filosofskie problemy geologicheskogo vremeni (Philosophical problems of the geological time concept): *In* Voprosy Filosofii, No. 2, Akad. Nauk SSSR, Inst. Filosofii, Izd. "Pravda", Moscow, p. 96–104.

Kosygin, Yu. A., Yu. S. Salin, and **V. A. Solovyev** (Editors), 1974, Stratigrafiya i matematika (Stratigraphy and mathematics): Akad. Nauk SSSR, Dalnevostochnyi nauchnyi tsentr, Inst. Tektoniki i Geofiziki, Khabarovsk, 207 p.

Kottlowski, F. E., 1958, Formation or formation—Discussion of Report 4 (of Am. Com. Strat. Nomen.) on Nature, Usage and Nomenclature of Rock Units: Am. Assoc. Petroleum Geol. Bull., v. 42, no. 4, p. 893–894.

Kovalevskiy, O. P., 1969, O granitsakh geologicheskikh sistem (The boundaries of geologic systems): *in* Problemy Stratigrafi, Trudy SNIGGIMSa, no. 94, p. 131–137, Novosibirsk. (English translation by Israel Program for Scientific Translations, *in* Classification in Stratigraphy, publ. for U.S. Dept. of Int. and Nat. Sci. Foundation, p. 116–121, Jerusalem, 1971.

———, 1971, Analiz osnovnykh zamechaniy k pravilam stratigraficheskoy klassifikatsii i terminologii (Analysis of main criticisms of rules of stratigraphic classification and terminology): Sov. Geol., no. 2, p. 43–55. (English translation *in* Int. Geol. Rev., v. 14, no. 3, 1972, p. 205–213.)

* **Krasheninnikov, V. A.,** 1969, Geograficheskoe i stratigraficheskoe raspredelenie planktonnykh foraminifer v otlozheniiakh paleogena tropicheskoĭ i subtropicheskoĭ oblasteĭ (Geographic and stratigraphic distribution of planktonic foraminifera in Paleogene deposits in tropical and subtropical regions): Trudy Geologich, in-ta Akad. Nauk SSSR, vyp. 202.

———, 1971, Cenozoic foraminifera: Initial reports of Deep Sea Drilling Project, v. 6, p. 1055–1068, Washington.

Krashennikov, G. F., 1968, O ponimanii termina "fatsiya" i yego geneticheskom soderzhanii (Meaning of "facies" and genetic substance of term): Bull. Moskovskogo obshchestva ispytateley prirody, otdel geologicheskiy, no. 2, p. 3–15. (English translation *in* Int. Geol. Rev., v. 10, no. 12, 1968, p. 1337–1345.

* **Krasnov, I. I.,** 1961, Recent state and further tasks of mapping and working out of stratigraphic nomenclature of the Quaternary deposits in the USSR (Title translated from Russian): *in* Materialy Vsesoiuznogo sovescania po izucheniiu chetvertichnogo perioda, Moskva, Izd. Akad. Nauk SSSR, v. 1, p. 89–98.

* ———, 1971, Problemy razrabotki stratigraficheskikh podrazdeleniĭ dlĭa detal'nogo raschleneniia antropogena (Problems in developing a stratigraphic classification for detailed subdivision of the Anthropogene): *in* Problemy periodizatsii pleĭstotsena. Materialy simpoziuma, Leningrad.

Krasnov, V. I., and **A. P. Shcheglov,** 1969, O raschlenenie tolshch v zavisimosti ot tectonicheskikh uslovii ikh formiroveniya (Classification of rocks according to the tectonic conditions of their formation): *in* Problemy Stratigrafi, Trudi SNIGGIMSa, no. 94, p. 121–130, Novosibirsk. (English translation by Israel Program for Scientific translations, *in* Classification in Stratigraphy, pub. Jerusalem, 1971, for U.S. Dept. of Int. and Nat. Sci. Foundation, p. 107–115.)

* **Krasny, L. I.,** 1952, On the question of stratigraphic nomenclature (Title translated from Russian): Trudi VSEGEI, Sbornik paleontologii i stratigrafii, p. 254–255.

Krassilov, V., 1974, Causal biostratigraphy: Lethaia, v. 7, no. 3, p. 173–179.

Krishtofovich, A. N., 1939, Novaiia sistema regional'noi stratigrafii (The new system of regional stratigraphy): Sov. Geol., v. 9, no. 9, p. 68–76.

———, 1946, Unifikatsiia geologischeskoi terminologii i novaia sistema regional'noi stratigrafii (Geological terminology and new system of regional stratigraphy): Paleontologischeskii Sbornik, no. 4, p. 46–76.

Krumbein, W. C., 1951, Some relations among sedimentation, stratigraphy, and seismic exploration: Am. Assoc. Petroleum Geol. Bull., v. 35, no. 7, p. 1505–1522.

——— and **L. L. Sloss,** 1951, Stratigraphy and sedimentation: W. H. Freeman & Co., San Francisco, Calif., 497 p.

——— and **L. L. Sloss,** 1963, Stratigraphy and sedimentation: 2nd ed., W. H. Freeman & Co., San Francisco, Calif., 660 p.

Krut, I. V., 1968, K sostoyaniyu ucheniya o geologicheskikh formatsiyakh (The state of the theory of geologic formations): Izv. Akad. Nauk SSSR, ser. geol., no. 9, p. 98–113. (English translation *in* Int. Geol. Rev., v. 11, no. 12, 1969, p. 1335–1346).

Krutzsch, W., and **D. Lotsch,** 1964, Contribution à la question de la subdivision du Tertiaire en deux systèmes indépendants: Le Paléogène et la "Néogène": BRGM France, Mém. 28 (Colloque sur le Paléogène, Bordeaux, 1962), p. 931–939.

* **Krymgol'ts, G. Ya.,** 1964, On the importance of some concepts in stratigraphy (Title translated from Russian): Trudi VSEGEI, novaia ser., v. 102, Obscieproblemy stratigrafii paleogena Turgaia i Srednei Azii, p. 20–24.

———, 1968, O nekotorykh kriteriyakh ustanovleniya stratigraficheskikh granits (Some criteria for the establishment of stratigraphic boundaries): Leningrad Univ., Vestn., no. 24, Geol. Geogr., no. 4, p. 175–176.

Kulp, J. L., 1960, Absolute age determination of sedimentary rocks: Fifth World Petroleum Cong. (New York, 1959), Proc., Section 1, p. 689–704.

———, 1960, The geological time scale: 21st Int. Geol. Cong. (Norden), pt. 3, p. 18–27.

——— (Conference ed.), 1961, Geochronology of rock systems: New York Acad. Sci., Ann., v. 91, Art. 2, p. 159–594. (A series of 57 papers by various authors resulting from a conference on Geochronology of Rock Systems held by the New York Acad. of Sciences, March 3–5, 1960).

———, 1961, Geologic time scale: Science, v. 133, no. 3459, p. 1105–1114.

Kummel, B., and **C. Teichert,** 1970, Stratigraphy and paleontology of the Permian-Triassic boundary beds, Salt Range and Trans-Indus Ranges, West Pakistan: *in*

Stratigraphic Boundary Problems: Permian and Triassic of West Pakistan (eds. B. Kummel and C. Teichert), Dept. Geol., Univ. Kansas, Spec. Pub. 4, p. 1–110.

Kutscher, F., 1960, Stratigraphische Tagesfragen: Notizblatt Hessischen Landesamt für Bodenforschung zu Wiesbaden, v. 88, p. 107–121.

Laffite, R., 1972, La notion stratigraphique d'étage: *in* Colloque sur les méthodes et tendances de la stratigraphie (Orsay, 1970), BRGM France, Mém. 77, pt. 1, p. 17–25.

—— **et al.,** 1972, Some international agreement on essentials of stratigraphy: Geol. Mag., v. 109, no. 1, p. 1–15.

—— **et al.,** 1972, Internationale Übereinkunft über die Grundlagen der Stratigraphie: Akad. Wiss. und Lit., Math-Naturwiss. Klasse, Abh., no. 1, 24 p.

—— **et al.,** 1972, Essai d'accord international sur les problèmes essentiels de la stratigraphie: Soc. Géol. France, C.R. Sommaire, fasc. 13, p. 36–45.

Lane, A. C., 1906, The geologic day: Jour. Geology, v. 14, no. 5, p. 425–429.

Lapparent, A. de, 1885, Traité de Géologie; Savy, Paris, 1504 p. (See p. 698–712, Principes de la classification des formations sédimentaires.)

Lapworth, C., 1879, On the tripartite classification of the Lower Paleozoic rocks: Geol. Mag., n.s., v. 6, p. 1–15.

Larson, R. L., and **T. W. C. Hilde,** 1975, A revised time scale of magnetic reversals for the Early Cretaceous and Late Jurassic: Jour. Geophys. Res., v. 80, no. 17, p. 2586–2594.

Lawson, A. C., 1930, The classification and correlation of the pre-Cambrian rocks: Calif. Univ. Dept. Geol. Bull., v. 19, no. 11, p. 275–293.

Lawson, J. D., 1962, Stratigraphical boundaries: Symposiums- Band, Silur/Devon-Grenze, Stuttgart (1960), p. 136–142.

——, 1971, The Silurian-Devonian boundary: Letter to Editor, Geol. Soc. London Jour., v. 127, pt. 6, p. 629–630.

——, 1971, Some problems and principles in the classification of the Silurian System: BRGM France, Mém. 73, p. 301–308.

——, 1971, Stratigraficheskiye printsipy i granitse Silura i Devona (Stratigraphic principles and the Silurian-Devonian boundary: Mezhdunar. Simp. Granitsa Silura Devona, Biostrat. Silura, Nizhego Srednego Devona, Tr., no. 3, v. 1, p. 135–144. (English summary).

Leckwijck, W. P. van, 1960, Report of the Subcommission of Carboniferous stratigraphy: C.R., 4th Cong. on Carboniferous, Heerlen, September 14-20, 1958, p. 24–26.

Lecompte, M., 1960, L'argument paléontologique en stratigraphie, quelques exemples critiques en Ardenne et dans l'Eifel: 21st Int. Géol. Cong. (Norden), pt. 21, p. 261–263.

——, 1962 (1961), Faciès marins et stratigraphie dans le Dévonien de l'Ardenne: Soc. Géol. Belgique Ann., t. 85, Bull. no. 1, p. B17–B57.

Le Conte, J., et al., 1898, A symposium on the classification and nomenclature of geologic time-divisions: Jour. Geology, v. 6, no. 4, p. 333–355.

Legrand, P., 1964, Considérations sur l'évolution de quelques concepts de stratigraphie (Application à l'exploration d'un nouveau bassin sédimentaire): BRGM France, Dept. Inform. Geol.. Bull Trimest., 16th year, no. 62, p. 1–8.

Lehmann, J. G., 1756, Versuch einer Geschichte von Flötz-Gebürgen, betreffend deren Entstehung, Lage, etc., Berlin, 76 p.

Leighton, M. M., 1958, Important elements in the classification of the Wisconsin glacial stage: Jour. Geology, v. 66, no. 3, p. 288–309.

————, 1958, Principles and viewpoints in formulating the stratigraphic classification of the Pleistocene: Jour. Geology, v. 66, no. 6, p. 700–709.

* **Leonov, G. P.,** 1953, On the question of principle and criteria of regional-stratigraphic subdivision of the sedimentary formations (Title translated from Russian): *in* Pamiati professora A.N. Mazarovicha, Moskva, p. 31–57.

————, 1955, On the question of correlation of stratigraphic and geochronological subdivisions (Title translated from Russian): Vestnik Moscovskogo Univ., no. 8, ser. geol., p. 17–31.

————, 1962, Problema tsiklichnosti v istorii zemli (The problem of cycles in geological history): Vest. Moskov. Un-ta., no. 4, p. 3–12. (English translation by M. E. Burgunker *in* Int. Geol. Rev., v. 6, no. 12, 1964, p. 2093–2099.)

*————, 1973, Osnovy stratigrafii (Fundamentals of stratigraphy): Tom. 1. Izd. Moskovsk, un-ta.

* ————, **V. P. Alimarina,** and **D. P. Naidin,** 1965, On principle and methods of establishment of stage subdivisions of standard scale (Title translated from Russian): Vestnik MGU, no. 4, ser. geol., p. 15–28.

LeRoy, L. W., 1951, Subsurface geologic methods, 2nd ed.: Colorado School of Mines, Golden, Colorado, 1156 p. (See p. 14–32.)

Librovich, L. S., 1948, Sur la méthode paléontologique en stratigraphie: Mat. VSEGEI, Paleont. i Strat. Sbornik 5. (Fide Stepanov as translated to French by Mme Jayet, S.I.G., Paris.)

———— **(ed),** 1954, Stratigraphic and geochronologic subdivisions (Title translated from Russian): Gosgeoltechizdat, 85 p., Moscow.

* ————, 1958, The lower boundary of the Carboniferous system and its designation (Title translated from Russian): Sov. Geol., no. 7, p. 35–42.

* ———— and **N. K. Ovechkin,** 1963, Tasks and rules of study and description of stratotypes and key stratigraphic sections (Title translated from Russian): MSK SSSR, VSEGEI, Moska, Gosgeoltehizdat.

Lloyd, A. J., 1964, The Luxembourg Colloquium and the revision of the stages of the Jurassic System: Geol. Mag., v. 101, no. 3, p. 249–259.

————, 1965, The Luxembourg Colloquium—a reply: Geol. Mag., v. 102, no. 1, p. 88.

Lochman-Balk, C., and **J. L. Wilson,** 1958, Cambrian biostratigraphy in North America: Jour. Paleontology, v. 32, no. 2, p. 312–350. (See particularly p. 313–318.)

Lohman, K. E. (for Committee on Jurisdictional Scope), 1963, Report 7 (of Am. Com. Strat. Nomen.)—Function and jurisdictional scope of the American Commission on Stratigraphic Nomenclature: Am. Assoc. Petroleum Geol. Bull., v. 47, no. 5, p. 853–855.

London, Geological Society of, 1967, Report of the Stratigraphical Code Sub-Committee "The Stratigraphic Code": Geol. Soc. London Proc., no. 1638, p. 75–87.

————, **The Royal Society,** 1968, International Geological Correlation Programme United Kingdom Contribution; Submitted to IUGS Prague, August 19–28, 1968: London, 43 p.

———, **Stratigraphy Committee,** 1969, Recommendations on stratigraphical usage: Geol. Soc. London Proc., no. 1656, p. 139–166.

Longacre, S. A., 1970, Trilobites of the upper Cambrian Ptychaspid biomere, Wilberns Formation, Central Texas: Jour. Paleontology, v. 44, pt. II of II, Supplement to No. 1, Mem. 4, p. 1–70. (See particularly p. 2–3.)

Lotze, F., 1952, Feinstratigraphische Studien I; Methodisches zur Feinstratigraphie des Turonpläners im Osning bei Lengerich: Neues Jahrb. Geol. Palaeontol., Mh., no. 10, p. 442–448.

Lowman, S. W., 1949, (Discussion) Sedimentary facies in geologic history: Geol. Soc. America Mem. 39, p. 145–151.

———, 1949, Sedimentary facies in Gulf Coast: Am. Assoc. Petroleum Geol. Bull., v. 33, no. 12, p. 1939–1997.

Lucas, J., and **M. Bonhomme,** 1972, Stratigraphie et géochronologie. Rapport sur les méthodes géochronologiques appliquées à la stratigraphie: *in* Colloque sur les méthodes et tendances de la stratigraphie (Orsay, 1970), BRGM France, Mém. 77, pt. 2, p. 936–941.

Lüttig, G., 1960, Vorschläge für eine geochronologische Gliederung des Holozäns in Europa: Eiszeitalter und Gegenwart, Bd. 11, p. 51–63.

———, 1964, Prinzipielles zur Quartär-Stratigraphie: Geol. Jahrb., v. 82, p. 177–202.

———, 1965, Interglacial and interstadial periods: Jour. Geology, v. 73, no. 4, p. 579–591.

———, 1967, Schleswig als Standardregion für die Internationale Holozän-Stratigraphie: *in* Frühe Menschheit und Umwelt, Teil II, p. 252–260.

———, 1968, Ansichten, Bestrebungen und Beschlüsse der Subkommission für Europäische Quartärstratigraphie der INQUA: Eiszeitalter und Gegenwart, Bd. 19, p. 283–288. Öhringen/Württ.

———, 1968, Bio-Zone, Chrono-Zone, Geo-Zone?: Geol. Jahrb., v. 86, p. 1–4, Hannover.

———, 1970, Comments against the stratigraphic use of the term "Villafranchium": Newsl. Strat., v. 1, no. 2, p. 61–66, Leiden.

———, 1970, Sprachlich-nomenklatorische Anregungen zur Unterscheidung von deutschsprachlichen Begriffen der Litho- und Ortho-Stratigraphie: Newsl. Strat., v. 1, no. 1, p. 53–58, Leiden.

———, 1970, Aktuelles zum Zwiegespräch Quartärgeologie—Urgeschichte: Nachrichten aus Niedersachsens Urgeschichte, Herausgegeben von der Archäologischen Kommission für Niedersachsen e.V., Bd. 39, Sonderdruck 1970, p. 17–27.

———, 1974, Die stratigraphische Kommission der Deutschen Union der geologischen Wissenschaften (DUGW): Newsl. Strat., v. 3, no. 2, p. 101–119.

——— **et al.,** 1969, Key to the interpretation and nomenclature of Quaternary stratigraphy: First and provisional edition, Hannover, 46 p. (multilith).

Lyell, C., 1833, Principles of geology, v. 3, Murray, London, 398 p. (plus 109 p.).

———, 1837, Principles of geology: Kay, Philadelphia, 2 vols., 1st American edition. (See particularly v. 2, p. 186–513.)

———, 1838, Elements of geology: Murray, London.

———, 1839, Elements of geology: French translation, appendix, p. 616–621, Paris.

MacGillavry, H. J., 1952, Wat is Stratigraphie: Rede uitgesproken bij de aanvaarding van het Ambt van Gewoon Hoogleraar aan de Universiteit van Amsterdam, 16 p.

Makedonov, A. V., 1968, Printsipy i metody regional'noi stratigrafii ugol'nykh basseinov, korrelyatsii razrezov i sinonimiki ugol'nykh plastov (Principles and methods of regional stratigraphy of coal basins, correlation of sections and synonymy of coal beds): *in* Metody korrelyatsii uglenosnykh tolshch i sinonimika ugol'nykh plastov, Moskva, Izd. "Nauka."

Malaysia, Geological Society of, 1968, Malaysian Code of Stratigraphic Nomenclature: Mimeographed sheets, 11 p.

Mallory, V. S., 1959, Lower Tertiary biostratigraphy of the California Coast Ranges: 416 p., Am. Assoc. Petroleum Geol., Tulsa, Oklahoma.

————, 1970, Biostratigraphy; a major basis for paleontologic correlation: *in* Correlation by Fossils, North Am. Paleontol. Conv. (1969), Proc., pt. F, p. 553–566.

Mamet, B., 1972, Quelques aspects de l'analyse sequentielle: *in* Colloque sur les méthodes et tendances de la stratigraphie, (Orsay, 1970), BRGM France, Mém. 77, pt. 2, p. 663–677.

Mangerud, J., S. T. Andersen, B. E. Berglund, and **J. J. Donner,** 1974, Quaternary stratigraphy of Norden, a proposal for terminology and classification: Boreas, v. 3, p. 109–128.

Mann, C. J., 1970, Isochronous, synchronous, and coetanecus: Jour. Geology, v. 78, p. 749–750.

Marcou, J., 1897, Rules and misrules in stratigraphic classification: Am. Geol., v. 19, no. 1, p. 35–49.

————, 1897, Rules and misrules in stratigraphic classification: Am. Geol., v. 19, no. 2, p. 111–131.

* **Markov, K. K.,** 1962, The main stratigraphic boundaries of the Quarternary system (Title translated from Russian): Trudi komissi po izucheniiu chetvertichnogo perioda Akad. Nauk SSSR, 20, p. 140–142.

Marr, J. E., 1898, The principles of stratigraphic geology: Cambridge Univ. Press, 304 p. (2nd ed., 1905).

Martin, W. D., and **B. R. Henniger,** 1969, Mather and Hockingport sandstone lentils (Pennsylvanian and Permian) of Dunkard Basin, Pennsylvania, West Virginia, and Ohio: Am. Assoc. Petroleum Geol. Bull., v. 53, no. 2, p. 279–298.

Martini, E., and **C. Müller,** 1971, Das marine Alttertiär in Deutschland und seine Einordnung in die Standard Nanoplankton Zonen: Erdöl und Kohle-Erdgas-Petrochemie vereinigt mit Brennstoff-Chemie, 24, Jahrg. no. 6, p. 381–384.

Martinsson, A., 1969, The series of the redefined Silurian System: Lethaia, v. 2, no. 2, p. 153–161.

————, 1973, Editor's column: Stratotypes: Lethaia, v. 6, no. 1, p. 101–102.

————, 1973, Editor's column: Cryptozoic and Phanerozoic: Lethaia, v. 6, no. 3, p. 311–312.

————, 1973, Editor's column: Ecostratigraphy: Lethaia, v. 6, no. 4, p. 441–443.

Martodjojo, S. (ed.), 1973, Sandi Stratigrafi Indonesia: Komisi Sandi Stratigrafi Indonesia, Ikatan Ahli Geologi Indonesia, 19 p. Revised edition, 1975, 19 p.

Martynov, V. A., 1969, Osobennosti stratifikatsi kontinental'nykh otiozhenii (Stratification characteristics of continental deposits): *in* Problemy Stratigrafi, Trudy SNIGGIMSa, no. 94, p. 150–155., Novosibirsk. (English translation by Israel Program for Scientific Translations, *in* Classification in Stratigraphy, pub. Jerusalem, 1971, for U.S. Dept. Int. and Nat. Sci. Foundation, p. 133–137.

* **Maslov, V. P.,** 1952, On stratigraphic subdivisions (Title translated from Russian): Izv. Akad. Nauk SSSR, ser. geol., no. 2, p. 140–141.

Mathews, A. A. L., 1949, Application of some biogenetic laws to stratigraphy: Texas Jour. Sci., v. 1, no. 3, p. 78–81.

Mattei, J., 1966, Méthode de corrélation biostratigraphique d'après des analyses d'associations de faunes d'invertébrés du Lias moyen et supérieur des Causses (Massif Central Français): Eclog. Geol. Helv., v. 59, no. 2, p. 916–926.

Maubeuge, P. L., 1959, Les méthodes modernes de la stratigraphie du Jurassique; ses buts, ses problèmes: Soc. Belg. Géol. Bull., v. 68, fasc. 1, p. 59–103.

———, 1963, La classification en stratigraphie et plus spécialement à la lumière du Jurassique méso-européan: *in* La Classification dans les Sciences: Soc. Belge de Logique et de Philosophie des Sciences, p. 89–116.

———, 1971, Le problème de la zone à *Arisphinctes plicatilis* de l'Oxfordien Moyen: un pas typique des problèmes de base de la biostratigraphie: Acad. et Soc. Lorraines des Sciences Bull., v. 10, no. 2, p. 77–100; Newsl. Strat., v. 2, no. 2, 1972, p. 97–113.

———, 1974, Observations à un essai de formulation des zones du Jurassique en France: Acad. et Soc. Lorraines des Sciences Bull., v. 13, no. 1, p. 23–37.

Maxey, G. B., 1964, Hydrostratigraphic units: Jour. Hydrology, v. 2, p. 124–129.

McCammon, R. B., 1970, On estimating the relative biostratigraphic value of fossils: Univ. of Uppsala Geol. Inst. Bull., new ser., v. 2, no. 6, p. 49–57.

McElhinny, M. W., and **P. J. Burek,** 1971, Mesozoic Palaeomagnetic stratigraphy: Nature, v. 232, no. 5306, p. 98–102.

McKee, E. D., 1943, Some stratigraphic principles illustrated by Paleozoic deposits of northern Arizona: Am. Jour. Sci., v. 241, no. 2, p. 101–108.

———, 1949, Facies changes in the Colorado plateau: Geol. Soc. America Mem. 39, p. 35–48.

———, and **C. W. Weir,** 1953, Terminology for stratification and cross-stratification in sedimentary rocks: Geol. Soc. America Bull., v. 64, no. 4, p. 381–389.

McKerrow, W. S., 1971, Palaeontological prospects—the use of fossils in stratigraphy: Geol. Soc. London Jour., v. 127, p. 455–464.

McLaren, D. J., 1959, The role of fossils in defining rock units with examples from the Devonian of western and arctic Canada: Am. Jour. Sci., v. 257, p. 734–751.

———, 1969, Report of IUGS Committee on the Silurian-Devonian Boundary and Stratigraphy, August 9, 1968: IUGS Geol. Newsletter, v. 1969, no. 1, p. 24–34.

———, 1970, Presidential address: time, life and boundaries: Jour. Paleontology, v. 44, no. 5, p. 801–815.

———, 1972, Report from the Committee on the Silurian-Devonian Boundary and Stratigraphy to the president of the Commission on Stratigraphy: IUGS Geol. Newsletter, v. 1972, no. 4, p. 268–288.

———, 1973, The Silurian-Devonian boundary: Geol. Mag., v. 110, no. 3, p. 302–303.

McLean, J. D., 1968, Foraminiferal zones and zone charts—an analysis and a compilation: *in* Manual of micropaleontological stratigraphy, v. 7, McLean Paleont. Lab., Alexandria, Virginia.

Melton, F. A., 1932, Time-equivalent versus lithologic extension of formations: Am. Assoc. Petroleum Geol. Bull., v. 16, no. 10, p. 1039–1043.

* **Menner, V. V.,** 1951, Principles of correlation of multifacial suites (Title translated

from Russian): Materialy paleontologicheskogo sovescaniia po paleozoiu, Moskva, p. 122–138.

* ——, 1960, On the question of nomenclature of the Upper Pre-Cambrian group (Title translated from Russian): Mejdunarodni geol. kong. 21st sessia, Doklady sov. geol., Problema 8, p. 201–207.

——, 1960, On the nomenclature problem of the upper Precambrian group: 21st Int. Geol. Cong. (Norden), pt. 8, p. 18–23.

* ——, 1961, Irregularity (stages) of evolution of organic world and its importance for detailed stratigraphy (Title translated from Russian): Trudy MGRI, v. 37, p. 177–183.

* ——, 1962, Biostratigraphic principles of correlation of marine, lagoon and continental suites (Title translated from Russian): Trudi geol. inst. Akad. Nauk SSSR, vypusk 65, p. 475.

——, 1969, General stratigraphic scale of Mesozoic and Cenozoic deposits in the USSR and the prospects for developing a single standard scale applicable in countries of the ECAFE region: in Stratigraphic correlation between sedimentary basins of the ECAFE region (3rd symposium on development of petroleum resources of Asia and the Far East, Tokyo, 1965), U.N. Econ. Comm. Asia Far East Mineral Resources Development Series, no. 30, p. 17–24.

——, 1971, Prostranstvennoye znacheniye stratigrafcheskikh podrazdeleniy (Geographical significance of stratigraphic subdivisions): Mosk. Ovshch. Ispyt. Prir. Byull., Otd. Geol., v. 46, no. 2, p. 9–16. (English translation in Int. Geol. Rev., v. 14, no. 1, 1972, p. 112–117.)

—— and V. A. Krasheninnikov, 1970, Contributions toward a stratigraphic scale for the ECAFE region: in Stratigraphic correlation between sedimentary basins of the ECAFE region, U.N. Econ. Comm. Asia Far East v. 2 of Mineral Resources Development Series no. 36, p. 19–25, United Nations, New York.

Merrill, W. M., 1958, Discussion of Note 19 (of Am. Com. Strat. Nomen.)—Status of soils in stratigraphic nomenclature: Am. Assoc. Petroleum Geol. Bull., v. 42, no. 8, p. 1978–1979.

Mesezhnikov, M. S., 1966, Zones of regional stratigraphic systems (Title translated from Russian): Sov. Geol., no. 7, p. 3–16.

——, 1969, Zonal'naya stratigrafiya i zoogeograficheskoye rayonirovaniye morsikh bassaynov (Biostratigraphy and distribution of faunal zones in sea basins): Geol. i Geofiz., (Akad. Nauk SSSR Sib. Otd.) no. 7, p. 45–53.

—— and V. N. Sachs, 1967, O sootnoshenii edinoy i regional'nykh stratigraficheskikh shkal (Relation of the universal and regional stratigraphic scales (pertaining to the article by S.G. Gurari and L.L. Khalfin on "Revision of the principles of stratigraphic classification is essential")): Geol. i Geofiz., no. 2, p. 145–147. (English translation in Int. Geol. Rev., v. 10, no. 1, 1968, p. 1–3.)

Middleton, G. V., 1973, Johannes Walther's Law of the correlation of facies: Geol. Soc. America Bull., v. 84, no. 3, p. 979–987.

Miller, T. G., 1964, The Luxembourg Colloquium: Geol. Mag., v. 101, no. 5, p. 469–471.

——, 1965, Time in stratigraphy: Paleontology, v. 8, pt. 1, p. 113–131.

Mintz, L. W., 1972, Historical geology—the science of a dynamic earth: Merrill, Columbus, Ohio, 785 p.

Mirsky, A., 1964, Reconsideration of the "Beacon" as a stratigraphic name in Antarctica:

in Antarctic Geology, First Int. Symposium, Capetown, 1963, Proc., p. 364–478, Interscience, New York.

Mitchum, R. M. Jr., P. R. Vail, and **J. B. Sangree,** 1974, Regional stratigraphic framework from seismic sequences (abstr.): Geol. Soc. America 1974 Ann. Meetings, abstracts with programs, v. 6, no. 7, October 1974, p. 873.

Momper, J. A., 1963, Nomenclature, lithofacies, and genesis of Permo-Pennsylvanian rocks—Northern Denver Basin: 1963 Rocky Mountain Assoc. of Geologists, p. 41–67.

———, 1966, Stratigraphic principles—with some applications to the Permo-Pennsylvanian of the Denver Basin: Wyoming Geol. Assoc., 20th Annual Conf., 1966, p. 90a–90r.

———, 1966, Stratigraphic principles applied to the study of the Permian and Pennsylvanian Systems in the Denver Basin: Wyoming Geol. Assoc. 20th Annual Conf., 1966, p. 87–89.

Monty, C. L. V., 1967, Pour une codification de la nomenclature stratigraphique Belge: Soc. Géol. Belgique Ann., v. 90, Bull. 3, p. B-203–253.

———, 1968, D'Orbigny's concepts of stage and zone: Jour. Paleontology, v. 42, no. 3, p. 689–701.

Moore, P. F., 1958, Nature, usage and definition of marker-defined vertically segregated rock units: Discussion: Am. Assoc. Petroleum Geol. Bull., v. 42, no. 2, p. 447–450.

Moore, R. C., 1936, Stratigraphic classification of the Pennsylvanian rocks of Kansas: State Geol. Survey of Kansas, Bull. 22, 256 p. (especially p. 29).

———, 1941, Stratigraphy: *in* Geology, 1888–1938: 59th Anniversary volume, Geol. Soc. America, p. 178–220.

———, 1948, Stratigraphical paleontology: Geol. Soc. America Bull., v. 59, no. 4, p. 301–326.

———, 1948, Classification of Pennsylvanian rocks in Iowa, Kansas, Missouri, Nebraska and northern Oklahoma: Am. Assoc. Petroleum Geol. Bull., v. 32, no. 11, p. 2011–2040.

———, 1949, Meaning of facies: Geol. Soc. America Mem. 39, p. 1–34.

———, 1950, Stratigraphical classification: Geol. Soc. Japan Jour., v. 56, no. 652, p. 39–47.

———, 1952, Orthography as a factor in stability of stratigraphical nomenclature: State Geol. Survey of Kansas, Bull. 96, pt. 9, p. 363–372.

———, 1952, Stratigraphical viewpoints in measurement of geologic time: Am. Geophys. Union Trans., v. 33, no. 2, p. 150–156.

———, 1955, Invertebrates and geologic time scale: Geol. Soc. America Special Paper 62, p. 547–574.

———, 1957, Minority report of ACSN Report-5—Nature, usage, and nomenclature of biostratigraphic units: Am. Assoc. Petroleum Geol. Bull., v. 41, no. 8, p. 1888.

———, 1958, Introduction to historical geology: McGraw Hill, New York, 656 p. (See p. 24–33.)

——— **et al.,** 1944, Correlation of Pennsylvanian formations of North America: Geol. Soc. America Bull., v. 55, no. 6, p. 657–706.

——— **et al.,** 1968, Developments, trends, and outlooks in paleontology: Jour. Paleontology, v. 42, no. 6, p. 1327–1377.

Morelock, J., and **O. Macsotay,** 1972, A review of the application of evolutionary concepts in stratigraphy: Lagena, no. 30, Inst. Oceanografico, Univ. de Oriente, Cumaná, Venezuela, p. 13–18.

Morley, L. W., and **A. Larochelle,** 1964, Paleomagnetism as a means of dating geological events: *in* Geochronology in Canada, p. 39–51, Royal Soc. Canada Spec. Pub. no. 8, Univ. Toronto Press, 156 p.

Morrison, R. B., 1967, Principles of soil stratigraphy: *in* Quaternary Soils (ed. by Morrison and Wright), 7th Cong. INQUA Proc., v. 9, p. 1–69.

———, 1968, Means of time-stratigraphic division and long-distance correlation of Quaternary succession: *in* Means of correlation of Quaternary successions (ed. by Morrison and Wright), Univ. of Utah Press, p. 1–113.

———, 1969, The Pleistocene-Holocene boundary: an evaluation, etc.: Geol. en Mijnbouw, v. 48, no. 4, pt. 2, p. 363–371.

* **Morosov, N. S.,** 1959, On some questions of construction of scheme of stratigraphic subdivisions (Title translated from Russian): Nauchnyi ejegodnik, Saratovskiy Univ., Geol. fakultat, p. 57–59.

Morton, N. (ed.), 1971, The definition of standard Jurassic stages: BRGM France, Mém. 75 (Colloque du Jurassique, Luxembourg, 1967), p. 83–93.

Moskvitin, A. I., 1959, Pleistocene stratigraphic and time units (Title translated from Russian): Akad. Nauk SSSR, Kom. Isucheniyu Chetvertich, Periods, B., no. 23, p. 3–16.

Mouterde, R., et al., 1971, Les zones du Jurassique en France: Soc. Géol. France C.R. somm. Séances, p. 76–102, Paris.

———, **C. Ruget,** and **B. Caloo,** 1972, Les limites d'étages. Examen du problème de la limite Aalenien-Bajocien: *in* Colloque sur les méthodes et tendances de la stratigraphie (Orsay, 1970), BRGM France, Mém. 77, pt. 1, p. 59–68.

Müller, A. H., 1951, Grundlagen der Biostratonomie: Deutschen Akad. Wiss. Berlin (Klasse Math. u. allgemeine Naturwiss.) Abh., Jahrg. 1950, no. 3, p. 1–147.

Muller, S. W., 1941, Standard of the Jurassic System: Geol. Soc. America Bull., v. 52, no. 9, p. 1427–1444.

———, 1961, Russian stratigraphic terms, stage names, and symbols: Int. Geol. Rev., v. 3, no. 3, p. 273–278.

——— and **H. G. Schenck,** 1943, Standard of Cretaceous System: Am. Assoc. Petroleum Geol. Bull., v. 27, no. 3, p. 262–278.

Munier-Chalmas, E., and **A. De Lapparent,** 1893, Note sur la nomenclature des terrains sédimentaires: Geol. Soc. France Bull., 3 sér., v. 21, p. 438–488.

Murchison, R. I., 1839, The Silurian System: London, 768 p. (Also see anonymous review in Edinburgh Review, April 1841, v. 147, p. 1–41.)

———, 1841, First sketch of some of the principal results of a second geological survey of Russia: London, Edinburgh and Dublin Philos. Mag. and Jour. Sci., 3rd ser., v. 19, no. 126, p. 417–422.

Murray, G. E., 1952, Vicksburg stage and Mosley Hill formation: Geol. Notes, Am. Assoc. Petroleum Geol. Bull., v. 36, no. 4, p. 700–707.

———, 1955, Midway stage, Sabine stage and Wilcox group: Am. Assoc. Petroleum Geol. Bull., v. 39, no. 5, p. 671–696.

——— and **L. J. Wilbert,** 1950, Jacksonian stage: Am. Assoc. Petroleum Geol. Bull., v. 34, no. 10, p. 1990–1997.

Nabholz, W. K., 1951, Beziehungen zwischen Fazies und Zeit: Eclog. Geol. Helv., v. 44, no. 1, p. 131–158.

Naidin, D. P., 1972 (Reviews of ISSC Reports 7a and 7b, in Russian): In Geologiya (Reference Journal), No. 12, p. 1–2.

Nalivkin, D. V., 1956, A study of facies: The environment of deposition (Title translated from Russian): Dept. Geol.-Geog. Sci. of USSR, Moscow, v. 1, 534 p., v. 2, 393 p. (Reviewed by M. Burgunker in Int. Geol. Rev., v. 1, no. 1, 1959, p. 103–111.

National Academy of Sciences (NAS) Division of Earth Sciences, 1967, Time and stratigraphy in the evolution of man: Publ. no. 1469, 97 p.

Nekhoroshev, V. P., 1970, O prichinakh, zatrudnyayushchikh korrelyatsiyu strati-graficheskikh skhem (Causes of difficulties in stratigraphic correlation): in Biostrati-graficheskiye i paleobiofatsial'nyye issledovaniya i ikh prakticheskoye znacheniye, p. 96–107, Vses. Paleontol. 0–vo, Moscow.

Nemtsovich, V. M., 1969, O vydelenii i klassificatsii geologicheskikh formatsiy (Identification and classification of geologic formations): Izv. Akad. Nauk SSSR, ser. geol., no. 10, p. 142–147. (English translation in Int. Geol. Rev., v. 12, no. 9, 1970, p. 1075–1079).

Neuman, R. B., and A. R. Palmer, 1956, Critique of Eocambrian and Infracambrian: in El Sistema Cámbrico, su paleogeografia y el problema de su base, pt. 1, 20th Int. Geol. Cong. (Mexico), p. 427–435.

Newell, N. D., 1961, Permo-Triassic hiatus in marine rocks of southeastern Europe: Am. Phil. Soc., Yearbook, p. 314–318.

————, 1962, Paleontological gaps and geochronology: Jour. Paleontology, v. 36, no. 3, p. 592–610.

————, 1962, Geology's time clock: Natural History, v. 71, p. 32–37.

————, 1963, Crises in the history of life: Sci. Am., v. 208, no. 2, p. 76–92.

————, 1966, Problems of geochronology: Acad. Nat. Sci., Philadelphia Proc., v. 118, no. 3, p. 63–89.

————, 1967, Paraconformities: in Essays in paleontology and stratigraphy (R.C. Moore commemorative volume): Univ. of Kansas, Dept. Geology Spec. Publ. no. 2, p. 349–367.

————, 1967, Revolutions in the history of life: Geol. Soc. America, Special Paper no. 89, p. 63–91.

————, 1971, Faunal extinction: in Encyclopedia of Science and Technology, 3rd ed., v. 5, McGraw Hill, New York, p. 208–210.

————, 1972, Stratigraphic gaps and chronostratigraphy: 24th Int. Geol. Cong. (Montreal), sec. 7, p. 198–204.

————, 1973, The very last moment of the Paleozoic Era: in The Permian and Triassic Systems and their mutual boundary, Canadian Soc. Petroleum Geol., Mem. no. 2, Calgary, (ed. by Logan and Hills), 766 p. (See p. 1–10.)

Newton, A. R., 1968, Correlation and nomenclature in the Precambrian: Geol. Soc. South Africa, Annexure to v. 71 for 1968, Rhodesian Branch, p. 215–224, Mardon Printers Ltd., Salisbury, Rhodesia.

New Zealand, Geological Society of, 1964, Questionnaire on stratigraphic nomenclature: 16 p.

————, 1965, Report of the subcommission to investigate the desirability of a New Zealand stratigraphic code: Geol. Soc. New Zealand Newsl., no. 18, 27 p.

————, 1967, Guide to stratigraphic nomenclature: Geol. Soc. New Zealand, 20 p.

Nichols, R. A. H., and **J. M. Wyman,** 1969, Interdigitation versus arbitrary cutoff: Resolution of an Upper Cretaceous stratigraphic problem, western Saskatchewan: Am. Assoc. Petroleum Geol. Bull., v. 53, no. 9, p. 1880–1893.

Nikolov, T., I. Sapunov, and **J. Stephanov,** 1965, Notes concerning the orthography of stage names: Bulgarian Geol. Soc. Rev., v. 26, no. 1, p. 115–117.

———— **et al.,** 1966, Lithostratigraphical units (nature, nomenclature, and classification): Bulgarian Geol. Soc. Rev., v. 27, pt. 3, p. 233–247. (In Bulgarian with English summary.)

Nishijima, S., 1970, Significance of the formational boundary from the point of view of petroleum geology (abstr.) (in Japanese): Geol. Soc. Japan Jour., v. 76, no. 2, p. 88.

North, F. J., 1931, Extracts from "From Giraldus Cambrensis to the geological map": Cardiff Naturalists Soc. Trans., v. 64, p. 20–29; 42–45; 50–53; 56–57; 66–69; 78–89; 89–97.

North, F. K., 1964, The geological time-scale: *in* Geochronology in Canada, p. 5–8, Royal Soc. Canada Spec. Pub. no. 8, Univ. Toronto Press, 156 p.

Norway, Commission on Stratigraphy of, 1961, Regler for Norsk stratigrafisk nomenklatur (Code of stratigraphical nomenclature for Norway): Norges Geol. Undersøkelse, no. 213, p. 224–233. (In Norwegian and English.)

Ogose, S., 1950, An opinion on the classification of strata: Geol. Soc. Japan Jour., v. 56, p. 459–469. (In Japanese with English abstract.)

————, 1953, On the stratigraphic nomenclature: Geol. Soc. Japan Jour., v. 59, p. 65–74. (In Japanese.)

Omalius d'Halloy, J. B. d', 1831, Eléments de Géologie: Levrault, Paris, 558 p. (See p. 79–97, De la division des terrains.)

————, 1831, Observations sur la classification des terrains: Soc. Géol. France Bull., v. 1 (1830–1831), 1st ser., no. 9, p. 213–220.

Oppel, A., 1856, 1. Die Juraformation Englands, Frankreichs und des südwestlichen Deutschlands, etc.: Jahreshefte Vereins für vaterlandische Naturkunde in Württemberg, Jahrg. 12 (II Aufsatze und Abhandlungen), p. 121–556, Verlag Ebner & Seubert, Stuttgart.

————, 1857, 3. Die Juraformation etc.: Jahreshefte Vereins für vaterlandische Naturkunde in Württemberg, Jahrg. 13, Ebner & Seubert, Stuttgart, p. 141–396.

————, 1858, Die Juraformation, etc.: Jahreshefte Vereins für vaterlandische Naturkunde in Württemberg, Jahrg. 14, Ebner & Seubert, Stuttgart, p. 129–291.

————, 1862, Palaeontologische Mittheilungen aus dem Museum des Koenigl. Bayer Staates: Stuttgart, 322 p.

Orbigny, A. d', 1850-52, Cours élémentaire de paléontologie et de géologie stratigraphiques: Masson, Paris, 2 vols.

Oriel, S. S., 1959, Problems of stratigraphic boundaries: See p. 5 of Paleotectonic Maps—Triassic System, Misc. Geol. Investigations. Map 1–300, U.S. Geol. Survey.

Orombelli, G., 1971, Concetti stratigrafici utilizzabili nello studio dei depositi: continentali quaternari: Riv. Ital. Paleont. Strat., v. 77, no. 2, p. 265–291.

* **Ovechkin, N. K.,** 1955, The All-Union meeting on general questions of stratigraphic classification and its results (VSEGEI, January 17–22, 1955) (Title translated from Russian): Sov. Geol., Sbornik 45, p. 161–173.

* ———, 1957, Some debate questions on stratigraphic classification (Title translated from Russian): Sov. Geol., Sbornik 55, p. 8–30.

* ———, 1957, The short review of activity of the Interdepartmental Stratigraphic Committee from June 1955 up to April 1957 (Title translated from Russian): Sov. Geol., Sbornik 58, p. 163–173.

* ———, 1961, The degree of study of stratigraphy of the territory of the USSR and further tasks (Title translated from Russian): VSEGEI Biull., no. 3, p. 5–24.

Owens, B., 1970, A review of the palynological methods employed in the correlation of Paleozoic sediments: Liège Univ., Cong. Colloq. (1969), v. 55, p. 99–112.

Oyen, F. H. van, 1964, La palynologie stratigraphique dans le cadre de la stratigraphie paléontologique: Inst. Français du Pètrole Rev., v. 19, no. 2, p. 183–195.

Paech, W., 1971, Zur Analyse des Begriffe der geologischen Formation: Zeit. Angew. Geol., v. 17, no. 5, p. 195–201.

Page, D., 1859, Classification of the materials composing the earth's crust into systems, groups and series: Chapter 6, p. 85–96 of Advanced text-book of geology, 2nd ed., 403 p.

Pakistan, Stratigraphic Nomenclature Committee of, 1962, Stratigraphic Code of Pakistan: Geol. Survey Pakistan Mem., v. 4, pt. 1, p. 1–8.

Palmer, A. R., 1954, The faunas of the Riley Formation in central Texas: Jour. Paleontology, v. 28, no. 6, p. 709–786.

———, 1965, Biomere—a new kind of biostratigraphic unit: Jour. Paleontology, v. 39, no. 1, p. 149–153.

Papp, A., F. Rögl, and J. Seneš, et al., 1973, Chronostratigraphie und Neostratotypen, Miozän der Zentralen Paratethys, Bd. III—M₂ Ottnangien die Innviertler, Salgotarjaner, Bantapusztaer Schichtengruppe und die Raehakia Formation: Vydavatel'stvo Slovenskej Akad. vied, Bratislava, 841 p.

Paproth, E., 1964, Die Untergrenze des Karbons: C.R., 5th Cong. Int. de Strat. et de Géol. du Carbonifère, Paris, September 9–12, 1963, p. 611–618.

Parker, F. L., 1965, Irregular distributions of planktonic foraminifera and stratigraphic correlation: in Progress in oceanography, v. 3, Pergamon Press, New York, p. 267–272.

Parks, J. M., Jr., 1953, Use of thermoluminescence of limestones in subsurface stratigraphy: Am. Assoc. Petroleum Geol. Bull., v. 37, no. 1, p. 125–142.

Patterson, J. R., and T. P. Storey, 1957, Lithologic versus stratigraphic concepts: Am. Assoc. Petroleum Geol. Bull., v. 41, no. 9, p. 2139–2142.

Payne, T. G., 1942, Stratigraphical analysis and environmental reconstruction: Am. Assoc. Petroleum Geol. Bull., v. 26, no. 11, p. 1697–1770.

Pearson, D. A. B., 1970, Problems of Rhaetian stratigraphy with special reference to the lower boundary of the stage: Geol. Soc. London Quart. Jour., v. 126, p. 125–150.

Pecherskiy, D. M., 1969, K voprosu ob odnovremennosti geologicheskikh protsessov (The synchronous nature of geologic processes): Izv. Akad. Nauk SSSR, ser. geol., no. 11, p. 110–115. (English translation in Int. Geol. Rev., v. 12, no. 9, 1970, p. 1107–1111.)

Perkins, R. D., 1974, Discontinuity surfaces as a stratigraphic tool: The Pleistocene of South Florida (abstr.): Geol. Soc. America 1974 Annual meetings—abstracts with programs, v. 6, no. 7, p. 908–909.

Perrodon, A., 1971, Des méthodes et des tendances de la stratigraphie: Newsl. Strat., v. 1, no. 4, p. 19–28, Leiden.

————, 1972, Conclusions et essai de synthese: *in* Colloque sur les méthodes et tendances de la stratigraphie (Orsay, 1970), BRGM France, Mém. 77, pt. 2, p. 985–999.

Phillips, J., 1840, Penny Cyclopaedia, v. 17, p. 153–154.

Pia, J. von, 1930, Grundbegriffe der Stratigraphie mit ausführlicher anwendung auf die europäische Mitteltrias: F. Deuticke, Leipzig-Vienna, 253 p.

————, 1937, Das Wesen der geologischen Chronologie: Deuxième Congrès pour l'avancement des études de stratigraphie carbonifère, Heerlen, September 1935, C.R., v. 2, p. 857–902.

Picard, M. D., 1960, Lithologic zone boundaries in Pennsylvanian Paradox Member, Paradox Basin: Am. Assoc. Petroleum Geol. Bull., v. 44, no. 9, p. 1574–1578.

Pichamuthu, C. S., 1970, On the use of the term "Archaean" in Precambrian stratigraphy: Curr. Sci. (India), v. 39, no. 23, p. 525–528.

Pletikapić, Z., 1969, Stratigrafija, paleogeograpfija, i naftoplinonosnost Ivanić-Grad formacija na obodu Moslavačkog massiya: GZH, Zagreb, 1969. (Fide Z. Boškov-Štajner).

Polovinkina, Yu. Ir., 1970, Sushchestvuyut li metamorficheskiye formatsii? (Do metamorphic formations exist?): *in* Regional·nyy metamorfizm i metamorfogennoye rudoobrazovaniye, p. 85–90, Akad. Nauk SSSR, Inst. Geol. Geokhronol. Dokembr., Leningrad.

Pomerol, C., 1969, Rapport sur la limite Paléocène-Éocène: BRGM France, Mém. 69, p. 447–449.

————, 1973, Stratigraphie et paléogéographie: Ère cénozoique: Doin, Paris, 272 p.

————, 1975, Stratigraphie et paléogéographie: Ère mésozoique: Doin, Paris, 384 p.

* **Poznamka, M.,** 1964, Mišika k návrhu slovenskej stratigrapfickej klassifikácie a terminologie: Geol. Sbornik, v. 15, no. 1, p. 172.

Powell, J. W., 1882, Plan of publication: U.S. Geol. Survey 2nd Ann. Report, 588 p. (See p. 40–54.), (In French *in* 2nd Int. Geol. Cong. (Bologna, 1881) C.R., p. 627–641.

————, 1888, Methods of geologic cartography in use by the United States Geological Survey: 3rd Int. Geol. Cong. (Berlin, 1885), C. R., p. 221–240.

————, 1890, Conference on map publication: U.S. Geol. Survey 10th Ann. Report, pt. 1, Geology, p. 56–79.

————, 1890, Nomenclature: U.S. Geol. Survey 10th Ann. Report, pt. 1, p. 63–67.

Prevot, M., 1972, Inversions de la polarité géomagnétique et stratigraphie: *in* Colloque sur les méthodes et tendances de la stratigraphie (Orsay, 1970), BRGM France, Mém. 77, pt. 2, p. 891–903.

Puillon-Boblaye, 1831, Notice sur les altérations des roches calcaires du littoral de la Grèce: Jour. de Géol., v. 3, p. 144–166.

Quennell, A. M., 1958, Nomenclature stratigraphique: Report of joint meeting of regional geological committees, Africa, in Leopoldville, Publ. no. 44, p. 11–26.

———— and **E. G. Haldemann,** 1960, On the subdivision of the Precambrian: 21st Int. Geol. Cong. (Norden), pt. 9, p. 170–178.

Quenstedt, W., 1951–52, Über grundlegende Begriffe der Stratigraphie und ihre Anwendung: Acta Albertina (Regensburger Naturwissenschaften), Bd. 20, no. 1, p. 47–52.

Raggatt, H. G., 1949, Australian Code of stratigraphic nomenclature: Mimeographed memorandum, 19 p., September 16, 1949.

———, 1950, Stratigraphic nomenclature: Australian Jour. Sci., v. 12, no. 5, p. 170–173.

———, 1953, A.N.Z.A.A.S. Standing Committee on stratigraphic nomenclature, first and second meetings: Australian Jour. Sci., v. 15, no. 4, p. 122–125.

———, 1956, Stratigraphical classification: correlation and identity: Australian Jour. Sci., v. 19, no. 1, p. 35.

———, 1956, Time division of Precambrian: Discussion: Am. Assoc. Petroleum Geol. Bull., v. 40, no. 2, p. 388.

———, 1957, Time division of Precambrian: Am. Assoc. Petroleum Geol. Bull., v. 41, no. 2, p. 333.

Rahman, H. (chairman), 1969, Report of the special working group on stratigraphic correlation between the sedimentary basins of the ECAFE region: *in* Stratigraphic correlation between sedimentary basins of the ECAFE region, U.N. Econ. Comm. Asia Far East, Tokyo meeting (1965), Mineral Resources Development Series, no. 30, p. 1–5.

———, 1969, Activities relating to stratigraphic correlation in Pakistan: *in* Stratigraphic correlation between sedimentary basins of the ECAFE region (3rd symposium on development of petroleum resources of Asia and the Far East, Tokyo, 1965), U.N. Econ. Comm. Asia Far East, Mineral Resources Development Series, no. 30, p. 13–16.

Rama Rao, L., 1964, The problem of the Cretaceous-Tertiary boundary with special reference to India and adjacent countries: Mysore Geologists' Assoc., 66 p.

———, 1968, Opening address: The problem of the Cretaceous-Tertiary boundary: *in* Cretaceous-Tertiary formations of South India (Seminar held at Bangalore, June 1966), Geol. Soc. India, Mem. no. 2, p. 1–9.

Rankama, K., 1967, Global Precambrian stratigraphy: Soc. Sci. Fennica Arsbok-Vuosikirja, 45B, no. 1, p. 1–14.

———, 1970, Global Precambrian stratigraphy: background and principles: Scientia, v. CV, no. DCIC-DCC, p. 382–421.

———, 1970, Proterozoic, Archean and other weeds in the Precambrian rock garden: Geol. Soc. Finland Bull., 42, p. 211–222.

———, 1973, A note on the terminology of ancient glaciogenic and similar nonglaciogenic sedimentary rocks and on their use in formal stratigraphic nomenclature: Geol. Survey New South Wales, Quart. Notes, October 1, 1973, p. 4–5.

Rastall, R. H., 1944, Palaeozoic, Mesozoic, and Kainozoic: a geological disaster: Geol. Mag., v. 81, p. 159–165.

Rat, P., 1972, Étude sur la zone et son emploi en stratigraphie: Comité Français de Stratigraphie, Compte-rendu de la séance de travail du 18 mars 1972 à Dijon, Feuille no. 4, 11 p.

* **Rausser-Chernoussova, D. M.,** 1953, Periodicity in the evolution of foraminifera of the Upper Paleozoic and its importance for subdivision and correlation of sections (Title translated from Russian): Materialy paleontologicheskogo sovescaniia po paleozoiu, Mai 14–17, 1951; Izd. Akad. Nauk SSSR, Moskva.

* ———, 1963, The historical evolution of fusulinids and the boundaries of stratigraphic subdivisions (Title translated from Russian): Voprosy mikropaleontologii, vypusk 7, p. 3–12.

————, 1966, Zur Frage des Zonenbegriffes in der Biostratigraphie: Eclog. Geol. Helv., v. 59, no. 1, p. 21–31.

————, 1967, O zonakh edinykh i regional·nykh stratigraficheskikh shkal (Zones in local and regional stratigraphic sequences): Izv. Akad. Nauk SSSR, ser. geol., no. 7.

Reeside, J. B. Jr., 1933, Stratigraphic nomenclature in the United States: 16th Int. Geol. Cong. (Washington), Guidebook 29, p. 1–7.

Reguant, S., 1971, Los conceptos de facies en estratigrafía: Acta Geol. Hispánica, t. 6, no. 4, p. 97–101.

————, 1973, Informacion y comentarios sobre la "Guia International de Clasificacion, Terminologia, y Usos Estratigraficos": Dept. Estrat. y Geol. Hist., Univ. Barcelona, no. 64, multilith, 5 p.

————, 1973, El Precámbrico: Dept. Estrat. y Geol. Hist., Univ. Barcelona, no. 63, multilith, 3 p.

————, **O. Riba,** and **A. Maldonado,** 1973, Acerca de los transitos verticales y horizontales en las secuencias estratigraficas: Dept. Estrat. y Geol. Hist., Univ. Barcelona, no. 65, multilith, 6 p.

* **Reiman, V. M.,** 1961, Stage subdivisions and stratigraphic nomenclature (Title translated from Russian): Izd. otd. geol.-himich. i technich. nauk Akad. Nauk Tadjikskoi SSSR, vypusk 3(5), p. 135–138.

Reiss, Z., 1966, Significance of stratigraphic categories—a review: Proc. 3rd sess. in Berne, Comm. Med. Neogene Strat. IUGS, June 8–13, 1964, p. 9–17 (E.J. Brill, publisher, Leiden).

————, 1968, Planktonic foraminiferids, stratotypes, and a reappraisal of Neogene chronostratigraphy in Israel: Israel Jour. Earth-Sci., v. 17, p. 153–169.

Remane, J., 1971, Les Calpionelles protozoaires planctoniques des mers Mésogéenes de l'époque secondaire: Annales Guébhard, 47 (1971, fasc. unique), p. 369–393.

Renevier, E., 1897, Chronographe géologique: 6th Int. Geol. Cong. (Zurich, 1894), p. 521–584. (See particularly Les facies ou formations, p. 528–581, and post-scriptum, p. 695).

————, 1901, Report of Commission Internationale du Classification Stratigraphique: 8th Int. Geol. Cong. (Paris, 1900), C.R., fasc. 1, p. 192–203.

Renier, A., 1952, Faunes et flores en stratigraphie de détail: 18th Int. Geol. Cong. (London), pt. 10, p. 5–9.

Renz, H. H., 1948, Stratigraphy and fauna of the Agua Salada Group, State of Falcon, Venezuela: Geol. Soc. America Mem. 32, 219 p.

Richarz, S., 1926, Biotic basis of stratigraphy: Pan-Am Geol., v. 46, p. 101–110.

Richmond, G. M., 1962, Discussion of Note 27 (of Am. Com. Strat. Nomen.)—Morphostratigraphic units in Pleistocene stratigraphy: Am. Assoc. Petroleum Geol. Bull., v. 46, no. 8, p. 1520–1521.

———— and **J. C. Frye,** 1957, Note 19 (of Am. Com. Strat. Nomen.)—Status of soils in stratigraphic nomenclature: Am. Assoc. Petroleum Geol. Bull., v. 41, no. 4, p. 758–763.

———— and **J. G. Fyles,** 1964, Note 30 (of Am. Com. Strat. Nomen.)—Application to American Commission on Stratigraphic Nomenclature for an amendment of Article 31, Remark (b) of the Code of Stratigraphic Nomenclature on misuse of the term "stage": Am. Assoc. Petroleum Geol. Bull., v. 48, no. 5, p. 710–711.

Richter, R., 1925, Über die Benennungsweise der Typen und über "offene Namgebung": Senckenbergiana, v. 7, p. 102–119.

——, 1937, Die unterscheidende Benennung von Stufe und Schicht nach der Weise von Siegen-Stufe und Siegener Schichten: Senckenbergiana, v. 19, no. 1–2, p. 116.

——, 1954, Die Priorität in der Stratigraphie und der fall Koblenzium/Siegenium/ Emsium: Senckenbergiana, v. 34, nos. 4–6, p. 327–338.

Riedel, W. R., 1973, Cenozoic planktonic micropaleontology and biostratigraphy: *in* v. 1 of Annual Review of Earth and Planetary Sciences (Annual Reviews Inc.), Palo Alto, Calif., p. 241–268.

——, **M. N. Bramlette,** and **F. L. Parker,** 1963, "Pliocene-Pleistocene" boundary in deep-sea sediments: Science, v. 140, no. 3572, p. 1238–1240.

—— and **A. Sanfilippo,** 1971, Cenozoic Radiolaria from the western tropical Pacific, Leg 7: Initial reports of the Deep Sea Drilling Project, v. 7, pt. 2, p. 1529–1672, Washington.

Rioult, M., 1969, Alcide d'Orbigny and the stages of the Jurassic: The Mercian Geologist, v. 3, no. 1, p. 1–30.

——, 1971, Alcide D'Orbigny et les étages du Jurassique: BRGM France, Mém. 75 (Colloque du Jurassique, Luxembourg, 1967), p. 17–33.

Rivera, R., 1956, Cronologia geologica clasica en el idioma castellano: Soc. Geol. Peru, Ann., pt. 1, v. 30, p. 329–333.

Rivero, F. C. de., 1965, Códigos estratigráficos: unos comentarios: Bol. Inform. Assoc. Venezolana Geol., Min. y Petroleo, v. 8, no. 8, p. 219–223.

Rivière, A., 1972, Place et rôle des méthodes paléoclimatiques en stratigraphie: *in* Colloque sur les méthodes et tendances de la stratigraphie (Orsay, 1970), BRGM France, Mém. 77, pt. 2, p. 699–703.

Roche, A., 1972, Faiblesses et possibilités de la méthode paléomagnétique dans son emploi en stratigraphie: *in* Colloque sur les méthodes et tendances de la stratigraphie (Orsay, 1970), BRGM France, Mém. 77, pt. 2, p. 853–855.

Rodgers, J., 1948, Note-6—Discussion of nature and classes of stratigraphic units: Am. Assoc. Petroleum Geol. Bull., v. 32, no. 3, p. 376–378, 380.

——, 1954, Nature, usage and nomenclature of stratigraphic units: a minority report: Am. Assoc. Petroleum Geol. Bull., v. 38, no. 4, p. 655–659.

——, 1959, The meaning of correlation: Am. Jour. Sci., v. 257, p. 684–691.

——, 1961, Time-stratigraphic boundaries of the Cambrian System and of its series (abstr.): 20th Int. Geol. Cong. (Mexico). (Publ. in Moscow), 2 p.

—— and **R. B. McConnell,** 1959, Note 23 (of Am. Com. Strat. Nomen.)—Need for rock-stratigraphic units larger than group: Am. Assoc. Petroleum Geol. Bull., v. 43, no. 8, p. 1971–1975.

Rodrigo, L. A., and **A. Castaños,** 1971, El problema del tiempo en geologia y el Código de Nomenclatura Estratigráfica: Yacimientos Pet. Fiscales Boliv., v. 1, no. 1, p. 58–66.

Roen, J. B., 1970, Sandstone distribution in Lower Member of Waynesburg Formation, Greene and Washington Counties, Southwest Pennsylvania: Discussion: Am. Assoc. Petroleum Geol. Bull., v. 54, no. 3, p. 532–534.

Roger, J., 1972, Vue d'ensemble sur les méthodes paléontologiques en stratigraphie: la biostratigraphie: *in* Colloque sur les méthodes et tendances de la stratigraphie (Orsay, 1970), BRGM France, Mém. 77, pt. 1, p. 449–457.

Rognon, P., 1972, Utilisation de certaines discontinuités sédimentaires d'origine climatique comme repères stratigraphiques: *in* Colloque sur les méthodes et tendances de la stratigraphie (Orsay, 1970), BRGM France, Mém. 77, pt. 2, p. 705–713.

Ross, C. A., 1970, Concepts in late Paleozoic correlations: Geol. Soc. America Special Paper 124, p. 7–36.

* **Rotay, A. P.,** 1953, Paleontological method and stratigraphy (Title translated from Russian): Materialy paleontologicheskogo sovescaniia po paleozoiu 1951, Izd. Akad. Nauk SSSR, Moscow, p. 88–91.

* ———, 1962, Paleontological method and problem of species in stratigraphy (Title translated from Russian): Izd. Kievskogo Univ., Kiev.

Rozanov, A. Yu, 1967, The Cambrian Lower boundary problem: Geol. Mag., v. 104, no. 5, p. 30, 416–434.

* ———, 1973, Zakonomernosti morfologicheskoĭ evoliutsii arkheotsiat i voprosy îarusnogo raschleneniia nizhnego kembriia (Laws controlling the morphologic evolution of the Archaeocyatha and problems of the subdivision of the lower Cambrian): Trudy Geologich, in-ta Akad. Nauk SSSR, vyp. 241.

Rubey, W. W., 1948, Discussion of nature and classes of stratigraphic units: Am. Assoc. Petroleum Geol. Bull., v. 32, no. 3, p. 378–380.

Rueller, K. H., 1971, Die Korrelation von Stratigraphie und physikalischen Daten der seismischen Aufnahmen, ein Versuch: Ingenieurmässige Lagerstättenbeschreibung, p. 273–296, Clausthal-Zellerfield.

Rutsch, R. F., 1939, Die Abtrennung des Paleocaens vom Eocaen: Eclog. Geol. Helv., v. 32, no. 2, p. 211–214.

———, 1952, Das typusprofil des Aquitanian: Eclogae Geol. Helvetiae, v. 44, (1951), no. 2, p. 352–355.

———, 1958, Das Typusprofil des Helvétien: Eclog. Geol. Helv., v. 51, no. 1, p. 107–118.

———, 1971, Helvetian: Giorn. Geol., Museo Geol. di Bologna Ann., ser. 2, v. 37 (1969), fasc. 2, p. 93–105.

——— and **R. Bertschy,** 1955, Der Typus des Néocomien: Eclog. Geol. Helv., v. 48, no. 2, p. 353–360.

Ryan, W. B. F., 1973, Paleomagnetic stratigraphy: Initial reports of the Deep Sea Drilling Project, v. 13, p. 1380–1387, Washington.

Sadykov, A. M., 1969, Sistema universal'noi stratigraficheskoi klassifikatsii (A System of universal stratigraphic classification): Izv. Akad. Nauk SSSR, ser. geol., no. 1

———, 1970, Zhacheniye i mesto stratigrafii v geologii (The significance and place of stratigraphy in geology): Akad. Nauk Kaz. SSSR, Izv., ser. geol., no. 5, p. 17–26.

Sando, W. J., et al., 1969, Carboniferous megafaunal and microfaunal zonation in the northern Cordillera of the United States: U.S. Geol. Survey Prof. Paper 613-E, 29 p.

Sanfilippo, A., and **W. R. Riedel,** 1973, Cenozoic radiolaria (exclusive of theoperids, artostrobids and amphipyndacids) from the Gulf of Mexico, Deep Sea Drilling Project Leg 10: Initial reports of the Deep Sea Drilling Project, v. 10, p. 475–611, Washington.

Sapunov, I. G., 1964, Notes on the boundary between the Lower and Middle Jurassic and on the stage term Aalenian: *in* Colloque du Jurassique, Luxembourg (1960). Volume des Comptes Rendus et Mémoires publié par l'Institut grand-ducal, Section des Sciences naturelles, physiques et mathématiques, p. 221–228.

Sartoni, S., and **U. Crescenti,** 1962, Ricerche biostratigrafiche nel Mesozoico dell'Appennino Meridionale: Giorn. Geol., ser. 2, v. 29 (1960–1961), p. 161–302.

Savage, D. E., 1973, Cenozoic—the primate episode: manuscript, 39 p.

Savitskiy, V. E., 1969, O pravilakh stratigraficheskoi klassifikatsii i terminologii i o prirode khronostratigraficheskikh podrazdelenii (Principles of stratigraphic classification and terminology and the nature of chronostratigraphic units): *in* Problemy Stratigrafi, Trudy SNIGGIMSa, no. 94, p. 84–99, Novosibirsk. (English translation by Israel Program for Scientific Translations, *in* Classification in Stratigraphy, pub. Jerusalem, 1971, for U.S. Dept. of Int. and Nat. Sci. Foundation, p. 73–86.)

————, 1969, Oyarusnom raschlenenii srednogo kembriya Sibiri i nekotorykh obshchikh voprosakh razrabotki etalonnoi shkaly yarusnykh podrazdelenni (The stage subdivision of the middle Cambrian and some general problems in elaborating a standard scale of stage units): *in* Problemy Stratigrafi, Trudy SNIGGIMSa, no. 94, p. 140–149, Novosibirsk. (English transl. by Israel Program for Scientific Translations, *in* Classification in Stratigraphy, pub. Jerusalem, 1971, for U.S. Dept. of Int. and Nat. Sci. Foundation, p. 124–132.)

————, 1970, O pravilakh opredeleniya nizhnev granitsy kembriya i granits drugikh krupnykh khronostratigraficheskikh podrazdelenii fanerozoya (Principles of determination of lower Cambrian boundary and boundaries of other major chronostratigraphic units of the Phanerozoic: *in* Materialy po regional'noy geologii Sibiri, Sib. Nauch-Issled Inst. Geol. Geofiz. Miner, Syr'ya, Tr., no. 110, p. 11–23.

Schatsky, N. S., 1960, Principles of late Precambrian stratigraphy and the scope of the Riphean Group: 21st Int. Geol. Cong. (Norden), pt. 8, p. 7–17.

Schaub, H., 1968, À propos de quelques étages du Paléocène et de l'Éocène du bassin de Paris et leur corrélation avec les étages de la Téthys: *in* Colloque sur l'Éocène, BRGM France, Mém. 58, p. 643–653.

Schenck, H. G., 1935, What is the Vaqueros Formation of California and is it Oligocene?: Am. Assoc. Petroleum Geol. Bull., v. 19, no. 4, p. 521–536.

————, 1940, Applied Paleontology: Am. Assoc. Petroleum Geol. Bull., v. 24, no. 10, p. 1752–1778.

————, 1961, Guiding principles in stratigraphy: Geol. Soc. India Jour., v. 2, 10 p.

————, **H. D. Hedberg,** and **R. M. Kleinpell,** 1935, Stage as a stratigraphic unit (abstr.): Pan-Am Geol., v. 64, no. 1, p. 70–71, August 1935; Geol. Soc. America Proc. for 1935, p. 347–348, June 1936.

———— and **R. M. Kleinpell,** 1936, Refugian stage of Pacific Coast Tertiary: Am. Assoc. Petroleum Geol. Bull., v. 20, no. 2, p. 215–225.

———— and **S. W. Muller,** 1936, Stratigraphic terminology (abstr.): Geol. Soc. America Proc. for 1935, p. 101–102 and p. 376, June 1936.

———— and **S. W. Muller,** 1937, Case-analyses of stratigraphic terms (abstr.): Geol. Soc. America Proc. for 1936, p. 296–297, June 1937.

———— **et al.,** 1941, Stratigraphic nomenclature—Discussion: Am. Assoc. Petroleum Geol. Bull., v. 25, no. 12, p. 2195–2212.

———— and **S. W. Muller,** 1941, Stratigraphic terminology: Geol. Soc. America Bull., v. 52, p. 1419–1426.

———— and **J. J. Graham,** 1960, Subdividing a geologic section: Sci. Rep. Tohoku Univ., 2nd ser. (Geol.) Spec. v. no. 4, "Professor Shoshiro Hanzawa Memorial Volume", p. 92–99.

Schindewolf, O. H., 1944, Grundlagen und Methoden der paläontologischen Chronologie: Gebrüder Borntraeger, Berlin-Zehlendorf, 139 p. (2nd ed., 1945.)

————, 1950, Grundlagen und Methoden der paläontologischen Chronologie: Berlin, 152 p., 3rd ed.

———, 1954, Über einige stratigraphische Grundbegriffe: Roemeriana, v. 1, Dahlgrün-Festschrift, p. 23–38. (English translation *in* Int. Geol. Rev., v. 1, no. 7, 1959, p. 62–70.)

———, 1954, Über die möglichen Ursachen der grossen erdgeschichtlichen Faunenschnitte: Neues Jahrb. Geol. Paläeontol., Monatsh., Bd. 10, p. 457–465.

———, 1955, Kleinforaminiferen und paläontologische Chronologie: Neues Jahrb. Geol. Paläont., Monatsh., Bd. 2, p. 82–84.

———, 1955, Die Entfaltung des Lebens im Rahmen der geologischen Zeit: Studium Generale, 8 Jahrg., H. 8, Springer-Verlag, Berlin, p. 489–497.

———, 1957, Comments on some stratigraphic terms: Am. Jour. Sci., v. 255, no. 6, p. 394–399.

———, 1958, Zur Aussprache über die grossen erdgeschichtlichen Faunenschnitte und ihre Verursachung: Neues Jahrb. Geol. Paläont., Monatsh., Bd. 6, p. 270–279.

———, 1960, Stratigraphische Methodik und Terminologie: Geol. Rundschau, v. 49, no. 1, p. 1–35.

———, 1963, Neokatastrophismus?: Deutsche Geol. Gesell. Zeit., Jahrg. 1962, v. 114, pt. 2, p. 430–445.

———, 1967, Logic and method of stratigraphy: Geol. Soc. South Africa, Trans., v. 67 (1964), p. 306–310.

———, 1970, Stratigraphical principles: Newsl. Strat., v. 1, no. 2, p. 17–24, Leiden.

———, 1970, Stratigraphie und Stratotypus: Akad. Wiss. Lit. Mainz, Math-Naturwiss. Kl., Abh., no. 2, 134 p.

Schmidt, H. E., and **E. Paproth,** 1972, Zur chronostratigraphischen Gliederung des Silesiums: Das Westfal mit besonderer Berücksichtigung Westfalens: C.R., Bd. 1, 7th Cong. Int. de Strat. et de Géol. du Carbonifère, Krefeld, August 1971, p. 209–211.

Schopf, J. M., 1960, Emphasis on Holotype (?): Science, v. 131, p. 1043.

Schuchert, C. E., 1916, Correlation and chronology in geology on the basis of paleogeography: Geol. Soc. America Bull., v. 27, p. 491–514.

———, 1937, What is the basis of stratigraphic chronology?: Am. Jour. Sci., v. 34, p. 475–479.

———, 1943, Stratigraphy of the eastern and central United States: Wiley, New York, 1013 p. (See p. 1–10, and 18–24.)

——— and **J. Barrell,** 1914, A revised geological time-table for N. America: Am. Jour. Sci., ser. 4, v. 38, no. 223, p. 1–27.

Scott, G. H., 1960, The type locality concept in time-stratigraphy: New Zealand Jour. Geol. Geophys., v. 3, no. 4, p. 580–584.

———, 1965, Homotaxial stratigraphy: New Zealand Jour. Geol. Geophys., v. 8, no. 5, p. 859–862.

———, 1967, Time in stratigraphy: New Zealand Jour. Geol. Geophys., v. 10, p. 300–301.

———, 1971, Revision of the Hutchinsonian, Awamoan, and Altonian stages (Lower Miocene, New Zealand)—1, New Zealand Jour. Geol. Geophys., v. 14, no. 4, p. 705–726.

Sdzuy, K., 1960, Zur Wende Präkambrium/Kambrium: Paläeont. Zeit., v. 34, p. 154–160.

———, 1962, Richtschnitt oder Leitfossil?: Symposium-Band, Silur-Devon Grenze, 1960, p. 231–233, Stuttgart.

Sedgwick, A., 1838, Synopsis of English series of stratified rocks . . . : Geol. Soc. London Proc., v. 2, no. 58, p. 675–690.

Seitz, O., 1931, Über Raum—und Zeitvorstellung in der Stratigraphie und deren Bedeutung für die stratigraphischen Grundprinzipien: Sitzungsber. Preuss. Geol. Landesamt., H. 6, p. 87–99, Berlin.

———, 1932, Ergänzende Bemerkungen über stratigraphische Raum und Zeitbegriffe: Jb. Preuss. Geol. Landesamt. für 1931, v. 52, p. 520–522, Berlin.

———, 1958, Gibt es eine Chronostratigraphie?: Geol. Jb., Bd. 75, p. 647–650.

Selli, R., 1960, Il Messiniano Mayer-Eymar 1867, proposta di un neostratotipo: Ann. Mus. Geol. Bologna, ser. 2, v. 28, p. 1–33.

———, 1967, The Pliocene-Pleistocene boundary in Italian marine sections and its relationship to continental stratigraphies: *in* Progress in Oceanography, v. 4, The Quaternary history of the ocean basins, Pergamon Press, p. 67–86.

——— (ed.), 1971 (1969), Stratotypes of Mediterranean Neogene stages: Giorn. Geol., Ann. Museo Geol. Bologna, ser. 2, v. 37, fasc. 2, p. 11–266. (A collection of reports on stratotypes of 14 Neogene stages by numerous authors.)

Semikhatov, M. A., 1966, K probleme obshchey stratigraficheskoy skhemy Dokembriye (Problem of a general stratigraphic scale for the Precambrian): Izv. Akad. Nauk SSSR, ser. geol., no. 4, p. 70–84. (English transl. *in* Int. Geol. Rev., v. 12, no. 4, 1970, p. 464–474.

Semikhatov, M. A., 1973, Obshchaya stratigraficheskaya shkala verkhnego dokembriya: sostoyanie i perspektivy (The general stratigraphic scale of the Upper Precambrian: the present state and perspectives): Akad. Nauk SSSR, ser. geol. no. 9, p. 3–17, Moscow.

Semikhatova, S. V., 1970, The nature of the biostratigraphic boundaries in the Lower Carboniferous deposits of the Russian platform (Title translated from Russian): Byul. Moskov, Obshchest. Ispytatelei Prirody, Otd. Geol., v. 45, no. 2, p. 104–116.

Seneš, J. (Sc. ed.), 1967, Chronostratigraphie und Neostratotypen, Miozän der Zentralen Paratethys, Bd. I—M₃ (Karpatien) Die Karpatische Serie und ihr Stratotypus (I. Chica, J. Seneš, J. Tejkal, et al.): Vydavatel'stvo Slovenskej Akad. vied Bratislava, 312. p. (See also under Cicha, I.)

——— (Sc. ed.), 1971, Chronostratigraphie und Neostratotypen, Miozän der Zentralen Paratethys, Bd. II—M₁ (Eggenburgien) Die Eggenburger Schichtengruppe und ihr Stratotypus (F. Steininger, J. Seneš, et al.). Vydavatel'stvo Slovenskej Akad. vied Bratislava, 827 p. (See also under Steininger.)

——— (Sc. ed.), 1973, Chronostratigraphie und Neostratotypen, Miozän der Zentralen Paratethys, Bd. III—M₂ Ottnangien (A. Papp, F. Rögl, and J. Seneš, et al). Vydavatel'stvo Slovenskej Akad. vied Bratislava, 841 p. (See also under Papp, A.)

Serra, O., 1972, Diagraphies et stratigraphie: *in* Colloque sur les méthodes et tendances de la stratigraphie (Orasy, 1970), BRGM France, Mém. 77, pt. 2, p. 775–832.

Shannon, J. P. Jr., 1962, Hunton group (Silurian-Devonian) and related strata in Oklahoma: Am. Assoc. Petroleum Geol. Bull., v. 46, no. 1, p. 1–29.

* **Shantser, E. V.,** 1960, Units of single and local stratigraphic scales of the Quaternary (Anthropogene) system. Draft of determination with respect to the North Eurasia (Title translated from Russian): Biull. MSK, no. 2, p. 61–64.

Shaver, R. H. (Geologic Names Committee of Indiana Geological Survey), 1962, Note 28 (of Am. Com. Strat. Nomen.)—Application to the American Commission on

Stratigraphic Nomenclature for an amendment of Article 4-f of the Code of Stratigraphic Nomenclature on informal status of named aquifers, oil sands, coal beds,
and quarry layers: Am. Assoc. Petroleum Geol. Bull., v. 46, no. 10, p. 1935.

────── (chairman, for **Geologic Names Committee of Indiana Geological Survey**), 1963,
Discussion of the stratigraphic code, beacon or gospel?: Am. Assoc. Petroleum Geol.
Bull., v. 47, No. 5, p. 850–851.

Shaw, A. B., 1964, Time in stratigraphy: McGraw-Hill, New York, 365 p.

Shimer, H. W., and **R. R. Shrock,** 1944, Index fossils of North America: Wiley, New
York, 837 p. (See p. 1–4.)

Shinbo, K., and **S. Maiya,** 1970, Neogene Tertiary planktonic foraminiferal zonation in
the oil-producing provinces of Japan: *in* Stratigraphic correlation between sedimentary basins of the ECAFE region, U.N. Econ. Comm. Asia Far East, Mineral Resources Development Series, no. 36 (1969), p. 135–142.

Sidorenko, A. V., and **V. A. Tenyakov,** 1972, Common historical and geological principles of investigation of Precambrian and Phanerozoic (abstr.): *in* Abstracts, 24th
Int. Geol. Cong. (Montreal), p. 28.

Sigal, J., 1961, Existe-t-il plusieurs stratigraphies?: BRGM France, Serv. Inform. Géol.,
Bull. Trimèst., 13th year, no. 51, p. 2–5.

──────, 1964, Une thérapeutique homéopathique en chronostratigraphie: les parastratotypes (ou prétendus tels): BRGM France, Serv. Inform. Geol., Bull. Trimest.,
no. 64, p. 1–8.

Silberling, N. J., and **E. T. Tozer,** 1968, Biostratigraphic classification of the marine
Triassic in North America: Geol. Soc. America, Spec. Paper no. 110, 63 p.

Simon, W., 1948, Zeitmarken der Erde: Grund und Grenze geologischer Forschung (Die
Wissenschaft, v. 98), Braunschweig Vieweg and Sohn, 232 p.

──────, 1960, Geologische Zeitrechnung in Dilemma: Naturwiss. Rundschau, Jahrg. 13,
H. 12, p. 461–465.

──────, 1962, Stratigraphische Gliederung, Terminologie, und Nomenklatur: *in* Leitfossi-
lien der Mikropaläontologie, Borntraeger, Berlin, p. 23–29.

────── and **H. J. Lippolt,** 1967, Geochronologie als Zeitgerüst der Phylogenie: *in* Die
Evolution der Organismen (ed. G. Heberer), v. 1, Fischer, Stuttgart, p. 161–237.

Simonson, R. W., 1952, Lessons from the first half century of soil survey: Pt. I, Classification of soils: Soil Sci., v. 74, p. 249–257.

──────, 1959, Soil classification and mapping in North Pacific Islands: 8th Pacific Science
Cong. Proc., v. 5, p. 149–162.

* **Sivov, A. G.,** 1955, On principles of building of regional stratigraphic scale and its
subdivisions (Title translated from Russian): Materialy Novosibirskoi konferencii po
ucheniiu o geologicheskih formaciiah., v. 1. Novosibirsk, p. 105–121.

Skinner, H. C., 1972, Gulf Coast stratigraphic correlation methods: Louisiana Heritage
Press, New Orleans, 213 p.

Sloss, L. L., 1958, Paleontologic and lithologic associations: Jour. Paleontology, v. 32, no.
4, p. 715–729.

──────, 1960, Concepts and applications of stratigraphic facies in North America: 21st
Int. Geol. Cong. (Norden), pt. 12, p. 7–18.

──────, 1960, Interregional time-stratigraphic correlation (abstr.): Geol. Soc. America
Bull., v. 71, no. 12, pt. 2, p. 1976.

──────, 1963, Sequences in the cratonic interior of North America: Geol. Soc. America
Bull., v. 74, no. 2, p. 93–114.

————, **W. C. Krumbein,** and **E. C. Dapples,** 1949, Integrated facies analysis: Geol. Soc. America Mem. 39, p. 91–123.

Smiley, T. L., 1964, On understanding geochronological time: Arizona Geol. Soc. Dig., v. 7, p. 1–12.

Smirnov, V. G., 1969, O neobkhodimosti reformy pravil stratigraficheskoi klassifikatsii (The need for a revision of the rules of stratigraphic classification): *in* Problemy Stratigrafi, Trudy SNIGGMSa, no. 94, p. 100–102, Novosibirsk. (English transl. by Israel Program for Scientific Translations, *in* Classification in Stratigraphy, pub. Jerusalem, 1971, for U.S. Dept. of Int. and Nat. Sci. Foundation, p. 87–89.)

Smith, J. P., 1901, The border line between Paleozoic and Mesozoic in western America: Jour. Geology, v. 9, no. 6, p. 512–521.

Smith, L. A., and **B. McNeely,** 1973, Summary of Leg 10, biostratigraphy: Initial reports of the Deep Sea Drilling Project, v. 10, p. 731–736, Washington.

Smith, W., 1815, Memoir to the map and delineation of the strata of England and Wales with a part of Scotland, Cary, London, 51 p.

————, 1816, Strata identified by organized fossils containing prints on coloured paper of the most characteristic specimens in each stratum: W. Arding, London, 32 p.

————, 1817, Stratigraphical system of organized fossils with reference to the specimens of the original collection in the British Museum explaining their state of preservation and their use in identifying the British strata, London.

Sokal, R. R., 1974, Classification: purposes, principles, progress, prospects: Science, v. 185, no. 4157, p. 1115–1123.

* **Sokolov, B. S.,** 1967, Pozdnii dokembrii i paleozoi Sibiri (mekotorye obshchie voprosy stratigrafii) (Late Precambrian and Paleozoic of Siberia (stratigraphic problems)): Geol. i Geofiz., no. 10.

————, 1968, Stratigraphic boundaries of Lower Paleozoic Systems: 23rd Int. Geol. Cong. (Prague), v. 9, p. 31–41.

* ————, 1971, Vend severa Evrazii (The Vendian of northern Eurasia): Geol. i Geofiz., no. 6.

————, 1971, Biokhronologiya i stratigraficheskiye granitsy (Biochronology and stratigraphical boundaries): *in* Problemy obschey i regional·noy geologii, p. 155–178, Akad. Nauk SSSR Sib. Otd., Inst. Geol. Geofiz., Novosibirsk.

Solle, G., 1962, Diskussion zur Silur/Devon-Grenze: Symposium Silur/Devon-Grenze, Bonn-Bruxelles (1960), ed. by H. K. Erben, 1962, p. 308–310, Stuttgart.

Solun, V. I., 1966, On methods of correlation of local and regional stratigraphic schemes (Title translated from Russian): Vestnik Leningrad Univ. Ser. Geol. i Geogr., no. 12, p. 23–29.

Sorgenfrei, Th., 1958, Molluscan assemblages from the Marine Middle Miocene of South Jutland and their environment: Geol. Survey of Denmark, 2 vols., 503 p., Copenhagen. (See particularly p. 11–19.)

South Africa, Geological Society of, 1971, South African Code of stratigraphic terminology and nomenclature: Geol. Soc. South Africa Trans., v. 74, p. 111–131.

Spieker, E. M., 1946, Late Mesozoic and Early Cenozoic history of Central Utah: U.S. Geol. Survey Prof. paper 205-D, p. 117–161.

————, 1949, Sedimentary facies and geologic structures in the Basin and Range Province: Geol. Soc. America Mem. 39, p. 55–82.

————, 1956, Mountain-building chronology and nature of geologic time scale: Am. Assoc. Petroleum Geol. Bull., v. 40, no. 8, p. 1769–1815.

Spizharskiy, T. N., 1963, O granitsye Kembriya i Dokembriya (The boundary between Cambrian and Precambrian): Sov. Geol., no. 8, p. 40–48. (English transl. *in* Int. Geol. Rev., v. 7, No. 8, 1965, p. 1368–1373.

Spry, A., and **M. R. Banks,** 1955, Stratigraphic nomenclature in the Precambrian: Australian Jour. Sci., v. 17, no. 6, p. 208–210.

Squires, D. F., 1960, Relative durations of the Tertiary Series and Stages in New Zealand: New Zealand Jour. Geol. Geophys., v. 3, no. 2, p. 137–140.

Stainforth, R. M., 1950, Is more concerted effort possible in establishing the regional significance of planktonic foraminifera as indices of geologic age?: Micropaleontologist (Am. Mus. Nat. Hist.), v. 4, no. 1, p. 18.

———, 1956, Meaning of the word stratigraphy: Discussion: Am. Assoc. Petroleum Geol. Bull., v. 40, no. 9, p. 2289–2290.

———, 1958, Stratigraphic concepts: Discussion: Am. Assoc. Petroleum Geol. Bull., v. 42, no. 1, p. 192.

Stamp, L. D., 1923, An introduction to stratigraphy (British Isles): London, 368 p.; 3rd ed. (1957), 381 p.

Stanton, T. W., 1930, Stratigraphic names: Report of Committee on Stratigraphic Nomenclature: Am. Assoc. Petroleum Geol. Bull., v. 14, no. 8, p. 1070–1079.

Steininger, F., J. Seneš, et al., 1971, Chronostratigraphie und Neostratotypen, Miozän der Zentralen Paratethys, Bd. II—M_1 (Eggenburgien) Die Eggenburger Schichtengruppe und ihr Stratotypus, Vydavatel'stvo Slovenskej Akad. vied Bratislava, 827 p.

Steinker, P. J., and **D. C. Steinker,** 1972, The meaning of facies in stratigraphy: The Compass, v. 49, no. 2, p. 45–53

Steno, N., 1669, De solido intra solidum naturaliter contento dissertationis prodomus: Florence, 76 p.

Stepanov, D. L., 1954, Methodologie des recherches stratigraphiques: Sputnik. Geologa-neftjanika, t. II. (Fide Stepanov as translated to French by Mme. Jayet, S.I.G., Paris.)

———, 1958, Principles and methods of biostratigraphic investigations (Title translated from Russian): Trudy VNIGRI, SSSR, no. 113, 180 p. (Transl. to French by Mme. Jayet, S.I.G., Trad. 2231, 148 p.)

———, 1967, Ob osnovnykh printsipakh stratigrafii (Basic principles of stratigraphy): Akad. Nauk SSSR, Izv. ser. geol., no. 10, p. 103–114.

* ———, 1969, Problemy stratigrafii verkhnego paleozoia (Stratigraphic problems of the upper Paleozoic): *in* Voprosy stratigrafii paleozoia, Izd. Leningradsk, un-ta.

———, **F. Golshani,** and **J. Stöcklin,** 1969, Upper Permian and Permian-Triassic boundary in North Iran: Geol. Survey of Iran, Report no. 12, 72 p.

Stephanov, J., 1966, The International Stratigraphic Scheme and the boundary between the Middle and Upper Jurassic: "Strasimir Dimitrov" Inst. Geol. Bull., v. 15, p. 79–88. (In English)

Stephenson, L. W., 1917, Tongue, a new stratigraphic term, with illustrations from the Mississippi Cretaceous: Washington Acad. Sci. Jour., v. 7, no. 9, p. 243–250.

Sterlin, B. P., M. S. Zinoviev, and **Ye. Migachera,** 1969, O podrazdeleniyakh obshchey i mestnoy stratigraficheskikh shkal (Regional and local divisions of the stratigraphic scale): Sov. Geol., no. 1, p. 38–44.

Stevenson, R. E., 1955, Two suggested rules for stratigraphic nomenclature: Geol. Notes, Am. Assoc. Petroleum Geol. Bull., v. 39, no. 12, p. 2524–2525.

Stipanicic, P. N., 1972, Kimmeridgiano (Piso) vs. Cimmerico (Diastrofismo): Assoc. Geol. Argent., Rev., v. 27, no. 2, p. 249–253.

Stirton, R. A., 1959, Time, life, and man: Wiley, New York, 558 p. (See p. 76–92.)

Stockwell, C. H., 1964, Principles of time-stratigraphic classification in the Precambrian: *in* Geochronology in Canada, Royal Soc. Canada Special Publication, no. 8, p. 52–60.

————, 1964, Fourth report on structural provinces, orogenies, and time-classification of rocks of the Canadian Precambrian shield: Geol. Survey Canada, paper 64–17, pt. II, p. 1–21.

————, 1973, Revised Precambrian time scale for the Canadian shield: Geol. Survey Canada, paper 72–52, 4 p.

———— and **R. K. Wanless,** 1961, Canadian shield age program of the Geological Survey of Canada: Geochronology of rock systems, New York Acad. Sci. Annals, v. 91, art. 2, p. 433–441.

Storey, T. P., and **J. R. Patterson,** 1959, Stratigraphy—traditional and modern concepts: Am. Jour. Sci., v. 257, p. 707–721.

Størmer, L., 1966, Concepts of stratigraphic classification and terminology: Earth Science Reviews, v. 1, no. 1, p. 5–28.

———— and **G. Henningsmoen,** 1960, Report of Committee on terminology of the Silurian and Ordovician: Mimeographed sheets, 6 p.

Straw, S. W., 1962, The Silurian-Devonian boundary in England and Wales: Symposium Silur/Devon-Grenze, Stuttgart 1960, p. 257–264.

Struve, W., 1966, Silurian [= Gotlandium] Statt Silur: Senckenbergiana Lethaia, v. 47, no. 2, p. 107–109.

Stubblefield, C. G., 1954, The relationship of paleontology to stratigraphy: Advancement of Science, no. 42, p. 149–159.

————, 1960, Reply to comments by P.A. Garrett: Nature, v. 187, no. 4740, p. 869.

Suggate, R. P., 1960, Time-stratigraphic subdivision of the Quaternary as viewed from New Zealand: Quaternaria, v. 5, p. 5–17.

Sun, Y. C., 1961, Problems of classification of the Cambrian System in China: Scientia Sinica, v. 10, no. 6, p. 726–733.

Suter, H. H., 1958, Note on the presentation of stratigraphic type sections: Alberta Soc. Petroleum Geol. Jour., v. 6, no. 1, p. 20–23.

Sutton, A. H., 1940, Time and stratigraphic terminology: Geol. Soc. America Bull., v. 51, no. 9, p. 1397–1412.

Suzuki, K., 1950, Critical review of the stratigraphical classification in recent years: Geol. Soc. Japan Jour., v. 56, p. 383–397. (In Japanese with English abstract).

Swann, D. H., and **H. B. Willman,** 1961, Megagroups in Illinois: Am. Assoc. Petroleum Geol. Bull., v. 45, no. 4, p. 471–483.

Sweet, W. C., H. Harper, Jr., and **D. Zlatkin,** 1974, The American-Upper Ordovician standard. XIX. A Middle and Upper Ordovician reference standard for the eastern Cincinnati region: Ohio Jour. Sci., v. 74, no. 1, p. 47–54.

Switzerland, Schweizerische Geologische Kommission, Arbeitsgruppe für Stratigraphische Terminologie, 1973, Empfehlungen zur Handhabung der stratigraphischen, insbesondere lithostratigraphischen Nomenklatur in der Schweiz: Eclog. Geol. Helv., v. 66, no. 2, p. 479–492.

Sylvester-Bradley, P. C., 1967, Towards an international code of stratigraphic nomenclature: *in* R. C. Moore Commemorative Volume (ed. by Teichert and Yochelson), Spec. Publ. no. 2, Dept. Geol., Univ. of Kansas, p. 49–56.

———, 1968, Hierarchy in stratigraphical nomenclature: Geol. Mag., v. 105, no. 1, p. 78.

Taiwan, China, Geological Society, 1970, Note on establishment of principles for standardization of stratigraphic nomenclature in Taiwan, China: *in* Stratigraphic correlation between sedimentary basins of the ECAFE region, U.N. Econ. Comm. Asia Far East, v. 2 of Mineral Resources Development Series, no. 36 (1969), p. 53.

Tedford, R. H., 1970, Principles and practices of mammalian geochronology in North America: *in* Correlation by fossils, North Am. Paleontol. Conv. (1969), Proc., pt. F, p. 666–703.

Teichert, C., 1950, Zone concept in stratigraphy: Am. Assoc. Petroleum Geol. Bull., v. 34, no. 7, p. 1585–1588.

———, 1957, Discussion of Report 5 (of Am. Com. Strat. Nomen.)—Nature, usage and nomenclature of biostratigraphic units: Am. Assoc. Petroleum Geol. Bull., v. 41, no. 11, p. 2574–2575.

———, 1958, Some biostratigraphical concepts: Geol. Soc. America Bull., v. 69, no. 1, p. 99–120.

———, 1958, Concepts of facies: Am. Assoc. Petroleum Geol. Bull., v. 42, no. 11, p. 2718–2744.

Termier, H., and **G. Termier,** 1964, Les temps fossilifères, I. Paléozoique Inférieur: Masson & Cie., Paris, 689 p. (See Introduction, p. 1–14).

Teslenko, Yu. V., 1969, K voprosu o vzaimootnoshenii edinoi i regional'nykh stratigraficheskikh shkal (The problem of the relationship between the general and regional stratigraphic scales): *in* Problemy Stratigrafi, Trudy SNIGGIMSa, no. 94, p. 79–83, Novosibirsk. (English transl. by Israel Program for Scientific Translations, *in* Classification in Stratigraphy, pub. Jerusalem, 1971, for U.S. Dept. Int. and Nat. Sci. Foundation, p. 68–72.

———, 1972, O kharaktere granits khronostratigraficheskikh podrazdeleniy yedinoy (Mezhdunarodnoy) stratigraficheskoy shkaly (Nature of the boundaries of chronostratigraphic units of the international stratigraphic scale): Geol. zhurnal (Russian ed.), v. 32, no. 3, p. 22–28.

* ———, 1974, Osnovnye polozheniia stratigrafii osadochnykh obrazovanii (Principles relating to the stratigraphy of sedimentary formations): Geol. zhurnal, v. 34, vyp. 1, Kiev.

Théobald, N., and **A. Gama,** 1959, Stratigraphie, Doin: Paris, 385 p. (See p. 7–24.)

Thomel, G., 1973, À propos de la zone à *Actinocamax plenus*: principe et application de la méthodeologie biostratigraphique: Ann. Mus. Hist. Nat. Nice, suppl. H.S., t. 1, p. 1–28.

———, 1973, De la méthode en biostratigraphie: Acad. Sci., Paris, C.R., v. 277, ser. D, p. 703–706.

Tintant, H., 1972, Paléontologie des invertébrés et stratigraphie: *in* Colloque sur les méthodes et tendances de la stratigraphie, (Orsay, 1970), BRGM France, Mém. 77, pt. 1, p. 33–39.

———, 1972, La conception biologique de l'espèce et son application en stratigraphie: *in* Colloque sur les méthodes et tendances de la stratigraphie (Orsay, 1970), BRGM France, Mém. 77, pt. 1, p. 77–87.

Tomlinson, C. W., 1940, Technique of stratigraphic nomenclature: Am. Assoc. Petroleum Geol. Bull., v. 24, no. 11, p. 2038–2048.

———, 1941, Reply to discussion by H. D. Hedberg of "Technique of stratigraphic nomenclature", by C. W. Tomlinson: Am. Assoc. Petroleum Geol. Bull., v. 25, no. 12, p. 2206–2207.

———, 1941, Reply to discussion by J. E. Eaton of "Technique of stratigraphic nomenclature", by C. W. Tomlinson: Am. Assoc. Petroleum Geol. Bull., v. 25, no. 12, p. 2210.

Torrens, H. S., 1971, Standard zones of the Bathonian: BRGM France, Mém. 75 (Colloque du Jurassique, Luxembourg, 1967, p. 581–604.

Tozer, E. T., 1965, Lower Triassic stages and ammonoid zones of Arctic Canada: Geol. Survey Canada, Dept. Mines and Technical Surveys, paper 65–12, 14 p.

———, 1967, A standard for Triassic time: Geol. Survey Canada, Bull. 156, 103 p.

———, 1971, Permian Triassic Boundary in West Pakistan: Geol. Mag., v. 108, no. 5, p. 451–455. (Essay review of Kummel, B. and C. Teichert, 1970, Stratigraphic boundary problems: Permian and Triassic of West Pakistan: Spec. Publ. no. 4, Univ. Kansas Press, 474 p.)

———, 1971, Triassic time and ammonoids: problems and proposals: Canadian Jour. Earth Sciences, v. 8, no. 8, p. 989–1031.

Trendall, A. F., 1966, Towards rationalism in Precambrian stratigraphy: Geol. Soc. Australia Jour., v. 13, pt. 2, p. 517–526.

Troelsen, J. C., and **Th. Sorgenfrei,** 1956, Principerne for stratigrafisk inddeling og nomenklatur (Procedure and terminology in stratigraphic classification): Dansk Geol. Foren., Meddel. Bd. 13, H. 3, p. 145–152.

Trowbridge, A. C., 1959, Stratigraphic Commission, Discussion of Report 6—Application of stratigraphic classification and nomenclature to the Quaternary: Am. Assoc. Petroleum Geol. Bull., v. 43, no. 3, p. 674–675.

Trueman, A. E., 1923, Some theoretical aspects of correlation: Proc. Geologists Assoc., v. 34, p. 193–206.

———, 1952, Some fundamental aspects of correlation (Summary): 18th Int. Geol. Cong. (London), pt. 10, p. 92–93.

Trümper, E., 1969, Zu einigen Problemen des Begriffes "Leitfossil": Deutsche Ges. Geol. Wiss., Ber., Reihe A, Geol. Paläontol., v. 14, no. 3, p. 349–355.

Trümpy, R., 1960, Über die Perm-Trias-Grenze in Ostgrönland und über die Problematik Stratigraphischer Grenzen: Geol. Rundschau, v. 49, p. 97–103.

Truswell, J. F., 1967, A critical review of stratigraphic terminology as applied in South Africa: Geol. Soc. South Africa Trans., v. 70, p. 81–116, Discussion, p. 189.

Turkey, Stratigraphic Committee of, 1968, Stratigrafi Siniflama ve Adlama Kurallari: Maden Tetlik ve Arama Enstitüsü Vayinlarindan, Ankara, 28 p.

Ulrich, E. O., 1916, Correlation by displacements of the strandline and the function and proper use of fossils in correlation: Geol. Soc. America Bull., v. 27, p. 451–490.

———, 1924, Notes on new names in table of formations and on physical evidence of breaks between Paleozoic Systems in Wisconsin: Wisconsin Acad. Sciences, Arts Letters, Trans., v. 21, p. 71–107.

Unesco, 1972, Intergovernmental conference of experts for preparing an International Geological Correlation Programme (IGCP), Paris, October 19–28 1971: Unesco, Paris, 52 p.

U.S. Geological Survey, Report by Committee, 1903, Nomenclature and classification for the geologic atlas of the United States: 24th Annual Report of the U.S. Geol. Survey for 1902 and 1903, p. 21–27.

————, 1953, Stratigraphic nomenclature in reports of the U.S. Geological Survey: U.S. Geol. Survey, Washington, D.C., 54 p.

U.S.S.R., All-Union Geological Inst. Paleont. and Stratigraphy, 1945, Uniformity in geological terminology and a new system of regional stratigraphy (Title translated from Russian): Materials All-Union Inst. Paleont. Strat., no. 4, p. 46–76.

U.S.S.R., 1950, Instructions on compilation of correlative stratigraphic schemes for the territory of the USSR and its separate regions: VSEGEI, Gosgeoltehizdat, Moskva.

U.S.S.R., Interdepartmental Stratigraphic Committee (A.P. Rotay, ed.), 1956, Stratigraficheskaya Klassifikatsiya i Terminologia (Stratigraphic classification and terminology): State Scientific-Technical Publishing House for Literature on Geology and Mineral Resources. (English transl. by John Rodgers, Int. Geol. Rev., v. 1, no. 2, 1959, p. 22–38.) (German transl. *in* Translations of Russian Geological and Paleontological Literature, no. 14, Inst. Paleont., Uppsala, 1960.)

———— **(A.P. Rotay, ed.),** 1960, Stratigraficheskaya Klassifikatsiya i Terminologia (Stratigraphic classification and terminology), 2nd revised ed.: State Publishing Office of Scientific and Technical Literature in Geology and Conservation of Mineral Wealth. In Russian, p. 5–31; in English, p. 32–58.

————, 1963, Zadachi i pravila izucheniya i opisaniya stratotipov i opornykh stratigraficheskikh razrezov (Problems and rules for study and description of stratotypes and stratigraphic reference cross-sections: State Geology and Mineral Conservation Engineering Press, Moscow. (English transl. by M. W. Burgunker, Int. Geol. Rev., v. 7, no. 7, 1965, p. 1141–1150.

———— **(A. I. Zhamoida, ed.),** 1965, Stratigraficheskaya Klassifikatsiya, Terminologia i Nomenklatura (Stratigraphic classification, terminology, and nomenclature): Izdatel'stvo Nedra Press, Leningrad, 70 p. (English transl., Int. Geol. Rev., v. 8, no. 10, pt. 2, 1966, p. 1–36.)

————, 1965, Decisions of Interdepartmental Stratigraphic Committee (1963) (regarding recommendations of first International Colloquium on Jurassic Systems): Int. Geol. Rev., v. 7, no. 5, 1965, p. 842–844.

* ————, 1965, Resolutions of the Interdepartmental Stratigraphic Committee and decisions of its permanent commissions on Paleogene and Quaternary deposits of the USSR (Title translated from Russian): Vypusk no. 6, Materialy conferencii, seminarov, sovescanii ONTI-VIEMS, Moskva.

* ————, 1965, Resolutions of the Interdepartmental Stratigraphic Committee and decisions of its permanent stratigraphic commissions on Lower Pre-Cambrian, Upper Pre-Cambrian, Ordovician and Silurian, Devonian, Triassic, Jurassic and Cretaceous of the USSR; Vypusk 7, Materialy conferencii, seminarov, sovescanii ONTI-VIEMS, Moskva.

———— **(A. I. Zhamoida et al.),** 1972, Main principles of the draft of the USSR stratigraphic code, report at the meeting of the International Subcommission on Stratigraphic Classification, Montreal, August 1972: Leningrad, 14 p. (In Russian and English.)

U.S.S.R., Standing Stratigraphic Commission on Paleogene of USSR, 1963, Resheniye postoyannoy Stratigraficheskoy Komissii MSK po Paleogeny SSSR (Decisions of the Standing Stratigraphic Committee on Paleogene of USSR): Sov. geol., 1963, no. 4, p. 145–154. (English transl., Int. Geol. Rev., v. 7, no. 4, 1965, p. 649–654.)

U.S.S.R., Standing Stratigraphic Commission on Ordovician and Silurian, 1963, Resheniye postoyannoy Stratigraficheskoy Komissii MSK po Ordoviskim i Siluriyskim Otlokheniyam SSSR (Decisions of the Standing Stratigraphic Commission on Ordovician and Silurian of USSR): Sov. Geol., no. 4, p. 141–144. (English transl., Int. Geol. Rev., v. 7, no. 4, 1965, p. 655–658.

Vail, P. R., and J. B. Sangree, 1971, Time stratigraphy from seismic data (abstr.): Am. Assoc. Petroleum Geol. Bull., v. 55, no. 2, p. 367–368.

——, R. M. Mitchum, and S. Thompson, III, 1974, Eustatic cycles based on sequences with coastal onlap: Geol. Soc. America 1974 Annual Meetings—abstracts with programs, v. 6, no. 7, p. 993.

Valentine, J. W., 1963, Biogeographic units as biostratigraphic units: Am. Assoc. Petroleum Geol. Bull., v. 47, no. 3, p. 457–466.

Vella, P., 1962, Biostratigraphy and paleoecology of Mauriceville District, New Zealand: Royal Soc. New Zealand Geol. Trans., v. 1, no. 12, p. 183–199.

——, 1964, Biostratigraphic units: New Zealand Jour. Geol. Geophys., v. 7, no. 3, p. 615–625.

——, 1965, Sedimentary cycles, correlation, and stratigraphic classification: Royal Soc. New Zealand Geol. Trans., v. 3, no. 1, p. 1–9.

Verwoerd, W. J., 1967, Stratigraphic classification: a critical review: Geol. Soc. South Africa Trans., 1964 (1967), v. 67, p. 263–282, (with discussions by A. R. Newton, J. F. Truswell, H. de la R. Winter, O.H. Schindewolf, H.D. Hedberg, and replies by W. J. Verwoerd, p. 304–316.

Vlerk, I. M. van der, 1959, Problems and principles of Tertiary and Quaternary stratigraphy: Geol. Soc. London Quart. Jour., v. 115, p. 49–64.

Votakh, O. A., 1972, Elementarnyye tektonicheskiye kompleksy zemnoy kory i geologicheskiye formatsii (Elemental tectonic complexes in the earth's crust and geologic formations): Izv. Akad. Nauk SSSR, Sov. Geol. i Geofiz., no. 8, p. 10–20. (English transl., Int. Geol. Rev., v. 15, no. 11, 1973, p. 1246–1254.)

Wagenbreth, O., 1965, Über Unschärfebeziehungen in der Geologie: Wiss. Z. Humboldt Univ., 4/5, p. 686–692.

——, 1966, Bemerkungen zum Zeitbegriff in der historischen Geologie und zur Frage einer Unschärfebeziehung bei rhythmischer oder zyklischer Schichtengliederung: Wiss. Zeit Hochschule für Architektur und Bauwesen Weimar, Jahrg. 13, no. 6, p. 617–625.

Wagner, R. H., 1969, Proposal for the recognition of a new "Cantabrian" Stage at the base of the Stephanian Series: 6th Int. Cong. on Stratigraphy and Geology of the Carboniferous, Sheffield, C. R., v. 1, p. 139–150.

——, F. J. Villegas, and F. Fonollá, 1969, Description of the lower Calabrian stratotype near Tejerina (Leon, N.W. Spain): 6th Int. Cong. on Stratigraphy and Geology of the Carboniferous, Sheffield, C. R., v. 1, p. 115–128.

Walcott, C. D., 1893, Geologic time as indicated by the sedimentary rocks of North America: Jour. Geology, v. 1, no. 7, p. 639–676.

——, 1903, Nomenclature and classification for the geological atlas of the United States: 24th Annual Report, U.S. Geol. Survey (1902–03), p. 21–27.

Wall, J. H., and R. K. Germundsen, 1963, Microfaunas, megafaunas, and rock-stratigraphic units in the Alberta group (Cretaceous) of the Rocky Mountain foothills: Canadian Petroleum Geol. Bull., v. 11, no. 4, p. 327–349.

Walliser, O. H., 1962, Diskussion zur Silur/Devon-Grenze: Symposium Silur/Devon-Grenze, Bonn-Bruxelles (1960) (ed. by H. K. Erben), Stuttgart, p. 311.

———, 1966, Die Silur/Devon-Grenze—Ein Beispiel biostratigraphischer Methodik: Neues Jahr. Geol. Palaeont. Abhandlungen, v. 125, p. 235–246.

Wang, Chao-Siang, 1964, In defense of traditional stratigraphy: Geol. Soc. China Proc., no. 7, p. 40–47.

———, 1973, Stratigraphic classification and terminology: an actualistic appraisal and proposal: Geol. Rundsch., v. 62, no. 3, p. 947–958.

Wang Hung, 1966, On rock-stratigraphic units: Acta Geol. Sinica, v. 46, no. 1, p. 1–13 (Abstr. in English, p. 12–13.)

Wanless, H. R., 1963, Termes stratigraphiques majeurs; Article Pennsylvanian: Lexique Strat. Int. Géol. Cong., Comm. Strat., Centre Nat. de la Recherche Scientifique, v. 8, 64 p.

Ward, L. K., 1952, Diastrophism and correlation: *in* Sir Douglas Mawson Anniversary Volume, Univ. Adelaide, p. 179–184.

Waterhouse, J. B., 1967, Proposal of series and stages for the Permian of New Zealand: Royal Soc. New Zealand Geol. Trans., v. 5, no. 6, p. 161–180.

———, 1968, New Zealand Permian stages: a rejoinder: New Zealand Jour. Geol. Geophys., v. 11, no. 1, p. 268–273.

———, 1969, Chronostratigraphy for the marine world Permian: Letter to the Editor, New Zealand Jour. Geol. Geophys., v. 12, no. 4, p. 842–848.

———, 1973, An Ophiceratid ammonoid from the New Zealand Permian and its implications for the Permian-Triassic boundary: Geol. Mag., v. 110, no. 4, p. 305–384.

Watkins, N. D., 1972, Review of the development of the geomagnetic polarity time scale and discussion of prospects for its finer definition: Geol. Soc. America Bull., v. 83, no. 1, p. 551–574.

———(?), 1973, Magnetic polarity time scale: Geotimes, v. 18, no. 5, p. 21–22.

Wedekind, R., 1916, Über die Grundlagen und Methoden der Biostratigraphie: Borntraeger, Berlin, 60 p.

———, 1918, Über Zonenfolge und Schichtenfolge: Zent. f. Min. Geol. u. Pal., Jahrg. 1918, p. 268–283.

Wegmann, E., 1962–1963, L'exposé original de la notion de faciès par A. Gressly (1814–1865): Sciences de la Terre, t. 9, no. 1, p. 83–119, Nancy.

Welin, E., 1966, The absolute time scale and the classification of Precambrian rocks in Sweden: Geol. Fören. Stockholm, Förhandl., v. 88, pt. 1, no. 524, p. 29–33.

Weller, J. M., 1958, Stratigraphic facies differentiation and nomenclature: Am. Assoc. Petroleum Geol. Bull., v. 42, no. 3, pt. 1, p. 609–639.

———, 1960, Stratigraphic principles and practice: Harper and Bros., New York, 725 p. (See p. 32–48.)

——— et al., 1948, Correlation of the Mississippian formations of North America: Geol. Soc. America Bull., v. 59, no. 2, p. 91–196. (See p. 124–128.)

Wells, J. W., 1963, Coral growth and geochronometry: Nature, v. 197, no. 4871, p. 948–950.

Wengerd, S. A., 1971 (1969), Chronostratigraphic analysis and the time surface: Soc. Geol. Mexicana Bol., v. 32, no. 1, p. 1–13.

Wezel, F. C., 1975, Diachronism of depositional and diastrophic events: Nature, v. 253, no. 5489, p. 255–257.

Wheeler, H. E., 1957, Le rôle de concepts stratigraphiques dans le problème de la frontière Cambrien-Précambrien: Colloque sur les relations Précambrien et Cambrien: Centre Nat. Recherche Sci. no. 76, Problème des series intermédiaires, Paris, 1957, p. 15–23.

————, 1958, Primary factors in biostratigraphy: Am. Assoc. Petroleum Geol. Bull., v. 42, no. 3, pt. 1, p. 640–655.

————, 1958, Time-stratigraphy: Am. Assoc. Petroleum Geol. Bull., v. 42, no. 5, p. 1047–1063.

————, 1959, Note 24 (of Am. Com. Strat. Nomen.)—Unconformity-bounded units in stratigraphy: Am. Assoc. Petroleum Geol. Bull., v. 43, no. 8, p. 1975–1977.

————, 1959, Stratigraphic units in space and time: Am. Jour. Sci., v. 257, p. 692–706.

————, 1963, Post-Sauk and Pre-Absaroka Paleozoic stratigraphic patterns in North America: Am. Assoc. Petroleum Geol. Bull., v. 47, no. 8, p. 1497–1526.

————, 1964, Baselevel, lithosphere surface, and time stratigraphy: Geol. Soc. America Bull., v. 75, no. 7, p. 599–610.

———— and **E. M. Beesley,** 1948, Critique of the time-stratigraphic connept: Geol. Soc. America Bull., v. 59, no. 1, p. 75–86.

———— **et al,** 1950, Stratigraphic classification: Am. Assoc. Petroleum Geol. Bull., v. 34, no. 12, p. 2361–2365.

———— and **V. S. Mallory,** 1953, Designation of stratigraphic units: Am. Assoc. Petroleum Geol. Bull., v. 37, no. 10, p. 2407–2421.

———— and **V. S. Mallory,** 1954, Analysis and classification of stratigraphic units (abstr.): Geol. Soc. America Bull., v. 65, no. 12, pt. 2, p. 1324.

———— and **V. S. Mallory,** 1956, Factors in lithostratigraphy: Am. Assoc. Petroleum Geol. Bull., v. 40, no. 11, p. 2711–2723.

Whewell, W., 1837, History of the inductive sciences: v. 3, Parker, London, 624 p. (See particularly sections on geological nomenclature and geological synonymy, p. 527–538.)

Whitaker, J. H. M., 1962, Diskussion zur Silur/Devon-Grenze: Symposium Silur/Devon-Grenze, Bonn-Bruxelles (1960) (ed. H.K. Erben), Stuttgart, 1962, p. 310–311.

White, C. A., 1894, The relation of biology to geological investigation: Report of U.S. Nat. Museum for 1892, p. 245–368. (See particularly p. 284–301.)

White, E. I., 1950, The vertebrate faunas of the lower Old Red Sandstone of the Welsh borders: British Mus. Nat. Hist (Geology) Bull., v. 1, p. 51–67.

White, R. T., 1940, Eocene Yokut sandstone north of Coalinga, California: Am. Assoc. Petroleum Geol. Bull., v. 24, no. 10, p. 1722–1751.

————, 1941, Discussion of "Technique of stratigraphic nomenclature" by C. W. Tomlinson: Am. Assoc. Petroleum Geol. Bull., v. 25, no. 12, p. 2210–2211.

Whiteman, A. J., 1970, Stratigraphic and Quaternary problems in West Africa: *in* Stratigraphy: an interdisciplinary symposium (ed. Daniels and Freeth), Ibadan Univ., Inst. African Studies Occas. Pub. no. 19, p. 29–33.

Whittington, H. B., and **A. Williams,** 1964, The Ordovician period. The Phanerozoic time-scale: A Symposium: Geol. Soc. London Quart. Jour., v. 120s, p. 241–254.

Wickman, F. E., 1948, Isotope ratios: A clue to the age of certain marine sediments: Jour. Geology, v. 56, no. 1, p. 61–66.

———, 1968, How to express time in geology: Am. Jour. Sci., v. 266, p. 316–318.

Wiedmann, J., 1967, Die Jura/Kreide-Grenze und Fragen stratigraphischer Nomenklatur: Neues Jahrb. Geol. Paläeontol, Monatsh., v. 12, p. 736–746.

———, 1968, Das Problem stratigraphischer Grenzziehung und die Jura/Kreide-Grenze: Eclog. Geol. Helv., v. 61, no. 2, p. 321–386.

———, 1970, Problems of stratigraphic classification and the definition of stratigraphic boundaries: Newsl. Strat., v. 1, no. 1, p. 35–48.

———, 1971, Die Jura/Kreide-Grenze Prioritäten, Diastrophen oder Faunenwende?: BRGM France, Mém. 75, (Colloque du Jurassique, Luxembourg, 1967), p. 333–338.

———, 1971, Problemas de la clasificación estratigráfica y de la definición de los límites estratigráficos: I Congreso Hispano-Luso-Americano de Geologia Economica, Madrid-Lisbon, 1971, v. 2, Sección 1, p. 785–805, Madrid.

———, 1972, Mass extinction on Mesozoic system boundaries (abstr.): Abstracts, 24th Int. Geol. Cong. (Montreal), p. 251.

———, 1973, Evolution or revolution of ammonoids at Mesozoic system boundaries: Biol. Reviews, Cambridge Philosophical Soc., v. 48, no. 2, p. 159–194.

Williams, H. S., 1891, Arkansas Geol. Survey Annual Report, v. 4, p. 13.

———, 1891, What is the Carboniferous System?: Geol. Soc. America Bull., v. 2, p. 16–19.

———, 1893, The making of the geological time scale: Jour. Geology, v. 1, no. 2, p. 180–197.

———, 1893, The elements of the geological time scale: Jour. Geology, v. 1, no. 3, p. 283–295.

———, 1894, Dual nomenclature in geological classification: Jour. Geology, v. 2, no. 2, p. 145–160.

———, 1895, Geological biology: Holt, New York, 395 p. (See p. 1–77.)

———, 1898, The classification of stratified rocks: Jour. Geology, v. 6, no. 7, p. 671–678.

———, 1901, The discrimination of time-values in geology: Jour. Geology, v. 9, no. 7, p. 570–585.

———, 1903, The correlation of geological faunas: U.S. Geol. Survey Bull. 210, 147 p.

———, 1905, Bearing of some new paleontologic facts on nomenclature and classification of sedimentary formations: Geol. Soc. America Bull., v. 16, no. 2, p. 137–150.

Williams, J. S., 1939, Lower Permian of type area, USSR: Washington Acad. Sci. Jour., v. 29, no. 8, p. 351–353.

———, 1951, Classification of Upper Paleozoic rocks in the United States: C.R., 3me Cong. Strat. Géol. du Carbonifère, Heerlen, p. 665–666.

———, 1954, Problem of boundaries between geologic systems: Am. Assoc. Petroleum Geol. Bull., v. 38, no. 7, p. 1602–1605.

——— and **A. T. Cross,** 1952, Note 13 (of Am. Com. Strat. Nomen.)—Third Congress of Carboniferous stratigraphy and geology: Am. Assoc. Petroleum Geol. Bull., v. 36, no. 1, p. 169–172.

Willis, B., 1901, Individuals of stratigraphic classification: Jour. Geology, v. 9, no. 7, p. 557–569.

———, 1912, Index to the stratigraphy of North America: U.S. Geol. Survey Prof. Paper 71, 894 p. (See p. 5–30.)

Willman, H. B., D. H. Swann, and J. C. Frye, 1958, Stratigraphic policy of the Illinois State Geological Survey: Illinois State Geol. Survey, Circular-249, 14 p., Urbana, Ill.

——— **and J. C. Frye,** 1970, Pleistocene stratigraphy of Illinois: Illinois State Geol. Survey Bull. 94, 204 p. (See p. 37–46.)

Wilmarth, M. G., 1925, The geologic time classification of the United States Geological Survey compared with other classifications; accompanied by the original definition of era, period, and epoch terms: U.S. Geol. Survey Bull. 769, 138 p.

Wilson, J. A., 1956, Miocene formations and vertebrate biostratigraphic units, Texas coastal plain: Am. Assoc. Petroleum Geol. Bull., v. 40, no. 9, p. 2233–2246.

———, 1959, Transfer, a synthesis of stratigraphic processes: Am. Assoc. Petroleum Geol. Bull., v. 43, no. 12, p. 2861–2865.

———, 1959, Stratigraphic concepts in vertebrate paleontology: Am. Jour. Sci., v. 257, no. 10, p. 770–778.

———, 1960, Stratigraphic practice in North American vertebrate paleontology: 21st Int. Geol. Cong. (Norden), pt. 22, p. 102–110.

———, 1971, Stratigraphy and classification: *in* Heinz-Tobien Festschrift, Abh. hess. L.-Amt Bodenforsch., p. 195–202.

Wilson, J. T., 1952, Some considerations regarding geochronology with special reference to Precambrian time: Am. Geophys. Union Trans., v. 33, no. 2, p. 195–203.

Winder, C. G., 1959, Contacts of sedimentary formations—a resume: Alberta Soc. Petroleum Geol. Jour., v. 7, no. 7, p. 149–156.

Woodford, A. O., 1963, Correlation by fossils: *in* The Fabric of Geology, (ed. C.C. Albritton), Addison-Wesley, New York p. 75–111.

———, 1965, Historical geology: W.H. Freeman & Co., San Francisco and London, 512 p. (See p. 153–190.)

Woodring, W. P., 1953, Stratigraphic classification and nomenclature: Am. Assoc. Petroleum Geol. Bull., v. 37, no. 5, p. 1081–1083.

Woodward, H. B., 1892, On geological zones: Geol. Assoc. Proc., v. 12, p. 295–315.

———, 1907, The history of the Geological Society of London: Geological Soc., London, 336 p. (See p. 18–24 on recommendations in 1808 for uniformity in geological nomenclature.)

Woodward, H. P., 1929, Standardization of geologic time-units: Pan-Am Geol., v. 51, no. 1, p. 15–22.

———, 1929, Priority in stratigraphic nomenclature: Science, n.s., v. 70, no. 1804, p. 96–97.

Worsley, T. R., and M. L. Jorgens, 1974, Automated biostratigraphy: manuscript, Dept. Oceanography, Univ. Washington, Seattle, 22 p.

Yakovlev, V. N., 1955, On the question of the scope of concepts of stratigraphic scale units (Title translated from Russian): Materialy Novosibirskoy konferencii po ucheniiu o geologicheskih formaciiah, v. 1, p. 123–129, Novosibirsk.

Yarkin, V. I., A. I. Zhamoida, et al., 1971, Osnovnyye polozheniya proyekta stratigraficheskogo kodeksa SSSR (Principal provisions of the project for a stratigraphic code in the USSR): Sov. Geol., no. 7, p. 47–55.

Yegoyan, V. I., 1969, O nekotorykh osnovnykh polozheniyakh obshchey stratigrafii (Fundamental positions of general stratigraphy): Izv. Akad. Nauk SSSR, ser. geol., no. 12, p. 3–13. (English transl., Int. Geol. Rev., v. 12, no. 10, 1970, p. 1206–1214.)

Yochelson, E. L., 1968, Biostratigraphy of the Phosphoria, Park City, and Shedhorn formations: U.S. Geol. Survey Prof. Paper 313-D, p. 571–660.

Young, K., 1959, Techniques of mollusc zonation in Texas Cretaceous: Am. Jour. Sci., v. 257, no. 10, p. 752–769.

———, 1960, Biostratigraphy and the new paleontology: Jour. Paleontology, v. 34, no. 2, p. 347–358.

Young, R. G., 1955, Sedimentary facies and intertonguing in the Upper Cretaceous of the Book Cliffs, Utah-Colorado: Geol. Soc. America Bull., v. 66, no. 2, p. 177–202.

Yuferev, O. V., 1969, Paleobiogeograficheskiye poyasa i podrazdeleniya yarusnoy shkaly (Paleobiogeographic zones and subdivisions of the stratigraphic scale): Izv. Akad. Nauk SSSR, ser. geol., no. 5, p. 77–84. (English transl., Int. Geol. Rev., v. 12, no. 5, p. 560–566.

———, 1969, Printsipy paleobiogeograficheskogo rayonirovaniya i podrazdeleniya stratigraficheskoy shkaly (Principles of paleobiogeographical regional classification and subdivision of the stratigraphic scale): Geol. i geofiz., no. 9, p. 19–28. (English transl., Int. Geol. Rev., v. 12, no. 8, 1970, p. 942–948.)

———, 1972, Stratigraficheskaya klassifikatsiya i terminologiya (Stratigraphic classification and terminology): Geol. i geofiz., Akad. Nauk SSSR, Sib. Otd., no. 1, p. 25–31.

Zagwijn, W. H., 1957, Vegetation, climate, and time-correlations in the early Pleistocene of Europe: Geol. Mijnbouw (n. ser.), Jaarg. 19, p. 233–244.

———, 1963, Pleistocene stratigraphy in the Netherlands, based on changes in vegetation and climate: Verhandelingen van het Koninklijk. Nederlands Geol. Mijnbouw. Genootschap. Geol. ser., deel 21–22, p. 173–196.

Zeiss, A., 1968 (1967), Untersuchungen zur Paläntologie der Cephalopoden des Unter-Tithon der Südlichen Frankenalb: Bayerischen Akad. Wissenschaften, Mathematisch-Naturwissenschaftliche Klasse, Abh. Neue Folge, v. 132, 190 p. (See particularly p. 127–133.)

Zeller, E. J., 1951, New determination of geologic age by the thermoluminescence method (abstr.): Geol. Soc. America Bull., v. 62, no. 12, p. 1493.

———, 1965, Modern methods for the measurement of geologic time: Mineral Inform. Serv., Kansas Univ., v. 18, no. 1, p. 12–15.

Zentralen Geologischen Institut der Deutschen Demokratischen Republik (ed.), 1968, Grundriss der Geologie der Deutschen Demokratischen Republik, Band 1 Geologische Entwicklung des Gesamtgebietes: Akad. Verlag, Berlin, 454 p.

Zeuner, F. E., 1952, Dating the past: an introduction to geochronology: 3rd revised ed., Methuan and Co. Ltd., London, 495 p.

Zhamoida, A. I., 1968, Sostoyanie i osnovnye zadachi stratigraficheskikh issledovanii v SSSR (The state and the main problems of stratigraphic investigations in the USSR): in Geologicheskoe stroenie SSSR, v. 1, (Stratigrafiya), Moskva, Izd. "Nedra".

———, 1969, Osnovnye voprosy stratigraficheskoi klassifikatsii, terminologii i nomenklatury (Principal problems of stratigraphic classification, terminology and nomenclature): in Geologicheskoe stroenie SSSR, t. 5, Izd. "Nedra".

——— (editor), 1970, Proekt stratigraficheskogo kodeksa SSSR (Scheme of a stratigraphic code for the USSR): Ministry of Geology of USSR—All-Union Scientific Research Inst. of Geol. (VSEGEI), Leningrad, 55 p. (English transl. by Israel Program for Scientific Translations, pub. Jerusalem, 1971, for U.S. Dept. Int. and Nat. Sci. Foundation, 36 p.)

——, O. P. Kovalevskiy and A. I. Moisseeva, 1969, Obzor zarubzhnykh stratigraficheskikh kodeksov (A survey of foreign stratigraphic codes): Interdepartmental Strat. Com. USSR Trans., v. 1, 103 p., Moscow. (English transl. by Israel Program for Scientific Translations, pub. Jerusalem, 1971, for U.S. Dept. Int. and Nat. Sci. Foundation, 72 p., under title "A survey of non-Soviet stratigraphic codes".)

—— et al, 1972, Main principles of the draft of the USSR stratigraphic code: Leningrad, 14 p. (In English and in Russian).

——, O. P. Kovalevskiy, A. I. Moisseeva, and V. I. Yarkin, 1973, Osnovnye diskussionnye voprosy po proektu stratigraficheskogo kodeksa SSSR (obzor zamechaniĭ) (Principal controversial problems of the project for a stratigraphic code for the USSR (summary of comments)): Postanovleniia Mezhvedomst. stratigr. komiteta i ego postoiannykh komissii, no. 13.

* ——, O. P. Kovalevskiy, V. V. Menner, A. I. Moisseeva, V. I. Yarkin, 1974, Osnovnye polozheniĭa proekta stratigraficheskogo kodeksa SSSR (Status of the project for a stratigraphic code for the USSR): Postanovleniia Mezhvedomst. stratigr. komiteta i ego postoiannykh komissi ĭi, vyp. 14.

——, O. P. Kovalevskiy, A. I. Moisseeva, and V. I. Yarkin (Compilers), 1974, Proekt stratigraficheskogo kodeksa SSSR: vtoroy variant (Project of a stratigraphic code for the USSR: second version): Ministerstvo Geologii SSSR, VSEGEI, Interdepartmental Stratigraphic Committee of USSR, Leningrad, 40 p.

—— and V. V. Menner, 1974, Dve osnoviye tendentsii razrabotki stratigrafichyeskoy klassifikatsii (Two fundamental principles of stratigraphic classification): in Problemi geologii i poleznikh iskopaemikh na xxiv sessii mezhdoonarodnolo geologichyeskolo kongressa, Akad. Nauk SSSR, p. 144–151, Moscow.

Zhemchuzhinikov, V. A., 1959, The problem of understanding the nomenclature of facies: Int. Geol. Rev., v. 1, no. 1, p. 65–72, (Transl. from Russian by E. Alexandroff.)

Zhizhchenko, B. P., 1958, Printsipy stratigrafii i unifitsirovannaia skhema deleniia Kainozoiska otlozheniy severnogo Kaukaza i smezhnykh oblastei (Principles of stratigraphy and unified scheme of subdivision of Cenozoic deposits of the North Caucasus and contiguous regions: Moscow, Gostoptechizdat, 312 p. (Transl. to French by M. Szyszman, Transl. no. 2242a, S.I.G., Paris, 1959.)

——, 1972, Kompleksnost'v reshenii voprosov stratigrafii kaynozoyskikh otlozheniy (A complex approach to the problem of stratigraphy of Cenozoic deposits): Sov. Geol., no. 2, p. 41–57.

Ziegler, B., 1971, Grenzen der Biostratigraphie im Jura und Gedanken zur stratigraphischen Methodik: BRGM France, Mém. 75, (Colloque du Jurassique, Luxembourg, 1967), p. 35–67.

—— and R. Trümpy, 1962, Sur les relations lithostratigraphiques entre le Rauracientype et l'Argovien-type: Colloque du Jurassique, 1962, Luxembourg, p. 293–300.

* Zinov·ev, M. S., E. E. Migacheva, and B. P. Sterlin, 1965, On volume, principles of establishment of zones and their correlation (Title translated from Russian): Sov. Geol., no. 5, p. 11–17.

Zubakov, V. A., 1963, Problema geologicheskoy sinkhronizatsii klimatostratigrafie (Problems of geological synchronization in climatostratigraphy: Sov. Geol., no. 8, p. 49–65.) (English transl., Int. Geol. Rev., v. 7, no. 8, 1965, p. 1374–1386.)

* ———, 1964, The critical survey of question on taxonomic rank of the Quaternary deposits (Title translated from Russian): Trudy Vsesoiuznogo Nauchno Issledovatelskogo Geol. Inst., v. 102, p. 80–103.

———, 1967, Paleontologic criteria of volume and rank of stratigraphic subdivisions: Int. Geol. Rev., v. 9, no. 11, p. 1415–1422. (Review-transl. of trans. of 8th session, All-Union Paleontological Soc. in 1962. Pub. by Nedra Press in Moscow, 1966.)

———, 1969, Diskussionnye voprosy stratigraficheskoi classifikatsii i terminologii (Controversial problems of stratigraphic classification and terminology): *in* Problemy Stratigrafi, Trudy SNIGGIMSa, no. 94, p. 43–65, Novosibirsk. (English transl. by Israel Program for Scientific Translations *in* Classification in Stratigraphy, pub. Jerusalem, 1971, for U.S. Dept. of Int. and Nat. Sci. Foundation, p. 35–44.)

———, 1969, Klassifikatsiya khronostratigraficheskikh podrazdeleniy klimaticheskogo soderzhaniya (Paleoclimatic classification of stratigraphic time units): Isv. Akad. Nauk SSSR, ser. geol., no. 1, p. 149–152.

* ———, 1973, Methodologic aspects of geochronology—stages and cycles as two interpretations of geologic time (Title translated from Russian): *in* Ritmichnost' prirodnykh iavlenii, Izd. "Nauka".

Zubkovich, M. E., 1960, On the importance of questions of stratigraphic classification (The study of Paleogene of Stalingrad Povolzh'e as an example) (Title translated from Russian): Trudy Vsesojuznogo proektnoizyskatelskogo i naucho-issledovatel·skogo instituta Gidroproekt sbornik 3, p. 234–241.

* ———, 1963, On principles of stratigraphic indexation (Title translated from Russian): Trudi vsesoiuznogo proektno-izyskatel-skogo i nauchno-issledovatel·skogo instituta Gidroproekt sbornik 9, p. 189–208.

* ———, 1968, Metody paleontologo-stratigraficheskikh issledovanii͡ (Methods of paleontologic and stratigraphic research): Osnovy biostratigrafii. Izd. "Vysshai͡a shkola".

Zudin, A. N., G. A. Pospelova, and **V. N. Saks,** 1969, Granitsy neogenovogo i chetvertichnogo periodov v svete paleomagnitnykh dannykh (Problem of the Neogene-Quaternary boundary in the light of paleomagnetic data): Geol. i Geofiz., no.͡ 8, p. 3–9. (English transl., Int. Geol. Rev., v. 12, no. 8, 1970, p. 989–994).

SOME PAPERS ON STRATIGRAPHIC CLASSIFICATION PUBLISHED IN RECENT YEARS IN THE PEOPLE'S REPUBLIC OF CHINA

(The following list is taken from A. I. Zhamoida, O. P. Kovalevskiy, and A. I. Moisseeva, 1969, Obzor zarubezhnykh stratigraficheskikh kodeksov (A survey of non-soviet stratigraphic codes): Interdepartmental Strat. Com. of USSR Trans., v. 1, 108 p., Moscow; transl. into English by Israel Program for Scientific Translations, Jerusalem, 1971, 72 p. (See p. 69–70), and published for the U.S. Dept. of Interior and Nat. Sci. Foundation, Washington.)

All-Chinese Stratigraphic Committee, 1960, The project of the stratigraphic code and its explanatory note (1st edition): Izdatel·stvo "Nauka".

———, 1963, The project of the stratigraphic code and its explanatory note (2nd edition): Izdatel·stvo "Nauka".

————, 1965, The project of the stratigraphic code and an explanatory note: Pekin, 54 p.

All-Chinese Stratigraphic Conference, 1959, Geochronological units, stratigraphic units, their indexing, and rules for subdivision of the regional stratigraphic scale (Project): Ti-Chih-Lun-ping (Geol. Review, v. 19 (5), p. 233–235.)

Chang Chia ch'i, 1959, A new variant of nomenclature rules for regional stratigraphic units: Ti-Chih-Lun-ping (Geol. Review, v. 19 (9), p. 432–433.)

Chang Shu-s'en, 1965, The problem of stratigraphic indexes: Ti-Chih-Lun-ping (Geol. Review, v. 23 (5), p. 392–303.)

Chao chung-fu, 1948, Some concepts on the usage of Chinese stratigraphic terms: Ti-Chih-Lun-ping (Geol. Review, v. 13, nos. 1–2.)

Chao Yi-yang, 1959, Unification of stratigraphic terminology: Ti-Chih-Lun-ping (Geol. Review), v. 19 (5), p. 229–230.)

————, 1959, Proposal of a new stratigraphic scale and its units: Ti-Chih-Lun-ping (Geol. Review, v. 19 (5), p. 230–232.)

Chou Wên-fu, 1964, Some problems of the "Project of a unified stratigraphic scheme": Nauchnyi Vestnik, no. 4, p. 364.

Hsieh Hsien-ming, 1959, Some considerations in connection with geochronologic and stratigraphic units: Ti-Chih-Lun-ping (Geol. Review, v. 19 (8), p. 381.)

Huang Pên-hung, 1959, Some considerations regarding geochronologic and stratigraphic units: Ti-Chih-Lun-ping (Geol. Review, v. 19 (10), p. 482–483.)

Mu En-chih, 1954, Stratigraphic terms: Ti-Chih-Chin-shin, v. 1, no. 1.

Shih T'ieh-min, 1959, The problem of stratigraphic units in the nomenclature of regional stratigraphic subdivisions: Ti-Chih-Lun-ping (Geol. Review, v. 19 (8), p. 380–381.)

Sun Yun-chou, 1961, Problems of classification of the Cambrian System of China: Ti-Chih-Hsüeh-pao (Acta Geol. Sinica, v. 41, (3/4), p. 285–289.)

Ting P'ei-ch'in, 1958, Applying stratigraphic nomenclature: Ti-Chih-Lun-ping (Geol. Review, v. 18 (3), p. 245–246.)

————, 1959, Concepts of the new stratigraphic code of China: Ti-Chih-Lun-ping (Geol. Review, v. 19 (9), p. 433–434.)

————, 1958, Application of chronostratigraphic units: Ti-Chih-Lun-ping (Geol. Review, v. 18 (3), p. 245.)

Wan Hung, 1966, Lithostratigraphic units: Ti-Chih-Hsüeh-pao (Acta Geol. Sinica, v. 46 (1), p. 1–12.)

Yang Hung-ta, 1957, Subdivision and naming of sediments: Izd. Nankinskogo universiteta.

Yeh-Lien-chün, 1960, Principles of classification and methods of subdivision of sedimentary facies and sedimentary formations: Ti-Chih-K'o-hsüeh, no. 2.

Yin Tsan-hsün, 1966a, Stratigraphic dictionary of China, No. 7. Carboniferous System, Review: Ti-Chih-Lun-ping (Geol. Review, v. 23 (4), p. 273.)

————, 1966b, The isolation and nomenclature of the largest stages in the development of the earth: Ti-Chih-Lun-ping (Geol. Review, v. 1 (1), p. 5.)

Index